John —

Thanks for all you've done for me over my career including the recent Royal Society stuff.

Jeff

Kinds, Things, and Stuff

NEW DIRECTIONS IN COGNITIVE SCIENCE

Series Editors
  Francis Jeffry Pelletier, Simon Fraser University
  Andrew Brook, Carleton University

Metarepresentations: A Multidisciplinary Perspective
  Edited by Dan Sperber

Common Sense, Reasoning, and Rationality
  Edited by Renée Elio

Reference: Interdisciplinary Perspectives
  Edited by Jeanette K. Gundel and Nancy Hedberg

Kinds, Things, and Stuff: Mass Terms and Generics
  Edited by Francis Jeffry Pelletier

Kinds, Things, and Stuff

Mass Terms and Generics

Edited by
Francis Jeffry Pelletier

UNIVERSITY PRESS

2010

OXFORD
UNIVERSITY PRESS

Oxford University Press, Inc., publishes works that further
Oxford University's objective of excellence
in research, scholarship, and education.

Oxford   New York
Auckland   Cape Town   Dar es Salaam   Hong Kong   Karachi
Kuala Lumpur   Madrid   Melbourne   Mexico City   Nairobi
New Delhi   Shanghai   Taipei   Toronto

With offices in
Argentina   Austria   Brazil   Chile   Czech Republic   France   Greece
Guatemala   Hungary   Italy   Japan   Poland   Portugal   Singapore
South Korea   Switzerland   Thailand   Turkey   Ukraine   Vietnam

Copyright © 2010 by Oxford University Press, Inc.

Published by Oxford University Press, Inc.
198 Madison Avenue, New York, New York 10016

www.oup.com

Oxford is a registered trademark of Oxford University Press.

All rights reserved. No part of this publication may be reproduced,
stored in a retrieval system, or transmitted, in any form or by any means,
electronic, mechanical, photocopying, recording, or otherwise,
without the prior permission of Oxford University Press.

Library of Congress Cataloging-in-Publication Data
Kinds, things, and stuff : mass terms and generics / edited by Francis
Jeffry Pelletier.
   p. cm. — (New directions in cognitive science)
Includes bibliographical references and index.
ISBN 978-0-19-538289-1
 1. Genericalness (Linguistics).   2. Semantics.   3. Grammar, Comparative
and general—Mass nouns.   4. Language and languages—Philosophy.
I. Pelletier, Francis Jeffry, 1944–
P299.G44K56 2009
415—dc22   2009021438

9 8 7 6 5 4 3 2 1

Printed in the United States of America
on acid-free paper

Preface

A very exciting event took place on February 3–4, 2006, at the downtown Vancouver campus (the "Harbour Centre") of Simon Fraser University. Ten scholars working at the intersection of philosophy of language, psychosemantics, linguistic theory, and child language development gathered to tell one another of the relevant research done in their home domains of scholarship. At the current distance in time, it is difficult to convey the sense of excitement that carried through this meeting. In addition to the usual presentation of scholarly works—where many of the participants were surprised to discover similarities of methods and conclusions, despite the different disciplines—there were two extended panel discussions: one about mass terms and the other about generics. These very lively discussions were marked by challenges to the methodologies of the participants and to their perceptions of the appropriate goals of their research. All of the scholars directly involved with the conference found it to be one of the most intellectually interesting conferences that they had ever been involved with, as did many of the audience members. Part of this enthusiasm was, of course, due to the actual talks given, whose written versions constitute this volume. But other forces that led to this excitement were the lively discussions that followed each talk and especially the panel discussions. The conference also featured nine posters contributed by students on a number of topics related to the theme of the conference. These posters provided an opportunity for the students and the speakers to engage in discussions, and this had the salutary effect of raising issues about how the narrower topics of generics and mass terms could shed light in the wider range of concerns that were present in the posters—and conversely, of course.

Acknowledgments

No international conference can be pulled off without the help of many sources and people. I first thank the speakers themselves for enthusiastically embracing my idea of a conference on cognitive aspects of mass terms and generics, and for willingly coming to rainy Vancouver in the middle of winter. Their talks and their contributions to the discussion as audience members, as well as their eager participation on related panel discussions and their serious engagement with the student posters, made this one of the most special conferences any of us can remember. And speaking of posters, the student contribution to the conference was quite special as well. All the traditional branches of cognitive science were represented by these students: anthropology, artificial intelligence, education, linguistics, philosophy, and psychology.

Financially, the conference benefited from many sources: Simon Fraser University's 40th Anniversary Fund (it was the 40th anniversary of SFU's foundation as a new public university in British Columbia!), SFU's Office of the Academic Vice President (John Waterhouse), the Faculty of Arts and Social Science (John Pierce), the Department of Linguistics, Department of Philosophy, Department of Psychology, Cognitive Science Program, and School of Computing Science. Additional funding was available through the University of British Columbia's Cognitive Systems program. The conference was opened by the president of Simon Fraser University, Michael Stevenson; further welcoming remarks were made by the director of cognitive science at Simon Fraser University, Fred Popowich; and the reception was ushered in by the associate dean of arts, Tom Perry.

I personally and especially thank Fred Popowich and Shamina Senaratne of the SFU Cognitive Science program for their tireless work in arranging both the strategic matters and the immediate matters of operating a conference. The SFU Cognitive Science Student Society provided the essential volunteers who did the work of registering participants and all that this entails. Finally, I am grateful to Greg Carlson for comments on, and suggestions for improvement to, the introduction.

Contents

Introduction   xi

PART I: Generics

1. Generics: A Philosophical Introduction   3
   Francis Jeffry Pelletier

2. Generics and Concepts   16
   Greg Carlson

3. Conceptual Representation and Some Forms of Genericity   36
   Sandeep Prasada

4. Are All Generics Created Equal?   60
   Francis Jeffry Pelletier

5. Stability in Concepts and Evaluating the Truth of Generic Statements   80
   James A. Hampton

6. Generics as a Window onto Young Children's Concepts   100
   Susan A. Gelman

PART II: Mass Terms

7. Mass Terms: A Philosophical Introduction   123
   Francis Jeffry Pelletier

8. A Piece of Cheese, a Grain of Sand: The Semantics of Mass Nouns and Unitizers   132
   Cliff Goddard

9. On Using Count Nouns, Mass Nouns, and *Pluralia Tantum*: What Counts?   166
   Edward J. Wisniewski

10. Count Nouns, Sortal Concepts, and the Nature of Early Words   191
    Fei Xu

Index   207

Introduction

Background

The topic of this volume is the cognitive aspects of generics and mass terms. This rather *recherché*-sounding topic actually has reflexes in many areas of research across the fields that are usually associated with cognitive science, as I will highlight in this general introduction and develop more fully in the introductory remarks that preface each of the two sections of the volume. As I am defining them here, both generics and mass terms are items of language. Most theorists, including all those in this volume, also think there are mental elements associated with these linguistic items, and that these elements could also be called "generic" or "mass." Also, many philosophers and linguists think there are aspects of the world that are associated with these linguistic items and that they, too, deserve to be called "generic" or "mass." One issue to be investigated is the relationships that might hold among these three different types of entities.

As linguistic items, generics are of two rather different types. One type is that of a noun or, more generally, an entire noun phrase, which purports to designate (or refer to) a *type*, or a *genus*. For example, the proper name *Ursus maritimus* and the noun phrase 'the polar bear' both refer to the species whose members are polar bears. Note that these linguistic items do not in themselves denote or refer to the individual polar bears, but rather to the *kind*. Such nouns and noun phrases are called "generic noun phrases" because they are taken to designate these (abstract) objects, genera. One question here concerns what sorts of noun phrases can designate genera (i.e., kinds). Some of the chapters in the generics section address this issue. The second type of generic is associated with entire sentences, namely, those sentences that express "general truths." For example, a sentence like 'Dogs have four legs' is thought to be true "generically," even though there are in fact some unfortunate dogs who have lost some legs, or maybe never had any. Generics of this type, the so-called "generic sentences," are thus distinguished from "particular sentences" like 'This dog has four legs' and also from explicitly quantified statements like 'Every dog has four legs'. A question of interest here concerns the types of predicates that can give rise to "generic truths" of this nature. Again, a number of chapters address this issue. Of course, there is also the issue of how any child, being raised in the physical and spatiotemporal world, could ever come to have generic knowledge. And this topic is also addressed in some of the chapters.

Linguistically, mass terms are nouns or, more generally, entire noun phrases, which are true of *stuff* rather than true of *things*. For example, *water* is true of stuff while *dog* is true of objects. This intuitive characterization is also associated with a number of other features that are used to distinguish mass from non-mass terms. For example, it seems that because a mass term is true of stuff, it will also be true of arbitrary parts of the stuff: take something that *water* is true of, divide it into two, and *water* will be true of both parts. On the contrary side, if we consider something that the noun *dog* is true of, divide it into two parts, we will discover that *dog* is not true of both parts. (Depending on how the division is carried out, it might not be true of either part!) In turn, this gives rise to the idea that non-mass words, or at least some of them, the so-called "sortal terms," have a natural and built-in "principle of counting," while mass terms do not. Thus, one can count how many dogs are in the room, but not how many waters are in the room (because all the parts of any water are also water, and so there is no number that characterizes it). Non-mass terms are normally called "count nouns," because they do allow for counting, and in so doing, they allow for pluralization—which mass terms do not.

All these features can occur together: 'Water is wet' uses the mass noun "water," but it uses it to refer to the *kind*, WATER (rather than to any specific instantiation of water). And it furthermore is a generic sentence, because 'is wet' is claimed to be generally true of water.

The explicit topic of the conference was *cognitive* issues involving generics and mass terms, indicating that special attention will be paid to how these linguistic items are related to the mental life of speakers who employ them. And the speakers all had things to say about this relationship. One intriguing issue pervading the conference was that of how the various pieces examined by the individual speakers fit together. For example, to what extent do mass terms versus count terms depend on the nature of the reality that language is used to describe, or do these encode different conceptualizations of the same thing? It is well known that a term might be grammatically count in one language and grammatically mass in another, for example, English mass 'hair' versus French count 'cheveux' (hairs). So, if one is speaking French and then switches to speaking English, need one reconceptualize the nature of hair itself, or does hair itself change to suit the language? And what of the alternatives offered in the same language, for example, count 'shoes' versus mass 'footwear,' and so on? Is there room here for a "natural language metaphysics" (Bach 1986), a level of ontological commitment that the semantics of a language commits one to by virtue of using a language, independently of the nature of reality or our individual conceptualizations of it?

As mentioned above, there were two very informative (as well as entertaining) panel discussions, one on generics and the other on mass terms. The panel members were the conference's speakers on the topic of the panel, and they were introduced and moderated by people whose contributions to the respective areas are substantial. And while it was perhaps initially thought that the panel sessions would be just a question–answer session on the already-presented papers, they turned into much, much more. The generics panel was moderated by John Lawler, who wrote the first book-length work on generic sentences to appear in the era of generative grammar and formal semantics (Lawler 1973b). This work, along with several contemporaneous articles (e.g., Lawler 1972, 1973a), established the agenda to be subsequently pursued by others in attempting to formulate, in general terms, a truth-conditional account of the semantics of this class of sentences. His observations, his characterizations of many of the problems that were encountered, and many of his examples are reflected in the substantial literature on the semantics of generics to this day. Lawler's commentary that interlaced the panel discussions presented a long and welcome perspective on the issues that the panel was, from varying points of view, grappling with.

The mass-term panel was moderated by Henry Laycock, who wrote influential philosophical articles in the early years of interest in the semantics (and ontology) of mass terms (see Laycock 1972, 1979). Laycock has recently published an entirely new work on the topic (Laycock 2006). Laycock opened the panel by making a pair of distinctions: singularity versus nonsingularity and plurality versus nonplurality. Using a number of examples, he illustrated ways the traditional count versus mass distinction may be inadequate, and challenged the panel members to show that any of their unusual or recalcitrant data weren't just due to the researcher's enforcing a simple distinction when there were really two distinctions at work. It was perhaps in this panel discussion that the two recurring themes of the conference discussions arose most strongly: the issue of "internalism versus externalism" in meaning, and the issue of natural language metaphysics. The former concerns the topic of what is the correct analysis of meaning for these sentences: does a sentence express a statement "about the world" and is made true or false because of how the world is organized, or does it gain its meaning from the mental life of the language users? Are the features associated with mass terms "in the world" or "in speakers' minds"? As can be seen, this is closely related to issues of natural language metaphysics: is there a separate study of "what a language ontologically commits its speakers to"? Needless to say, perhaps, is that there was no general agreement about answers to these these questions.

This Volume

Although there are natural areas where these different phenomena can occur together, there is also a natural division between generics and mass terms. This volume, like the conference, is thus divided in two, with the chapters that deal with generics forming one part, and those dealing with mass terms forming the other part. Each of the two parts starts with a philosophico-linguistic introduction to the sorts of issues that have made these two topics seem important to linguists, philosophers, artificial intelligence researchers, and other scholars, and indicating where cognitive research might shed some light on outstanding issues.

Generics

After the introductory essay that situates the study of generics in various fields, there are five chapters on different aspects of generics. In chapter 2, Greg Carlson points to the different uses of the terms "meaning" and "concept" in the literature that surrounds discussions of generic terms. It is usual in the formal semantics and philosophical literature on the topic to take a truth-conditional, "externalist" view as to what meaning is, while the psychological literature takes a mentalistic, "internalist" view. There is a similar conflict in how the term "concept" is used, with the philosophical literature favoring an "objective" sense (which might be cashed out independently of people, or maybe in terms of an interpersonal society in general) and the psychological and experimental literature employing a "subjective" sense of the term. Carlson considers what theorists might "want" from a theory of concepts, and proceeds to show the ways that differing desires here lead to different accounts of the phenomena surrounding generics, even when the underlying data seem to be more or less the same. In this regard, he considers work by Prasada and Dillingham (2006), Greenberg (2003), and Cohen (2005), and some unpublished work of Lyn Frazier and Charles Clifton, exposing them to a wider range of the types of generic sentences than is usually encountered. The general thrust of Carlson's efforts here are to try to narrow the divide between formal semantics of natural language and the experimental literature, especially on how they treat generics.

In chapter 3, Sandeep Prasada is concerned with the fact that although we encounter only a very limited number of instances of any given kind, we acquire generic knowledge. Although we have seen only a limited number of dogs, we acquire the knowledge 'Dogs have four legs'. What mechanisms make this happen? There are at least two preliminary background things that need discussing before this larger question can be answered: what is the nature of the knowledge acquired, and what are the available conceptual mechanisms that can be appealed to. Among the interesting conclusions suggested here is that there are reasons to prefer "generic representations" over "exemplar-based representations" for kinds. Prasada surveys the sorts of relations between the properties of the observed instances of a kind and the generic properties that are thereby attributed to that kind, invoking his distinction (see Prasada 2000, Prasada and Dillingham 2006) between k-properties of a kind and t-properties of a kind. In his parlance, k-properties are long term and quasi essential, whereas t-properties are transitory and accidental in nature. ('A train travels on tracks' attributes a k-property to trains, while 'A bicyclist is concerned with the environment' attributes a t-property to bicyclists—which might be "statistically true" but not reveal any "essential property".) A number of very interesting experiments are cited to illustrate the sort of picture that emerges.

Jeff Pelletier in chapter 4 focuses on the differing ways that generic sentences can be made. These sentences that report general characteristics are often given by using a sentence adverb modifier, such as 'usually' or 'typically' For instance, we might hear 'Birds typically fly' or 'Usually, birds fly'. But sometimes they are presented with adjectival modifiers such as 'normal' or 'typical,' as in 'The typical bird flies.' Additionally, the quantifier 'most' can be used, and the result called a generic statement: 'Most birds fly' And finally, there is the bare plural form: 'Birds fly'. Pelletier is interested in the question of whether these different forms give different information. For example, is the sentence 'Most birds fly' an "extensional" statement that compares the proportion of flying to nonflying birds? Is the sentence 'Birds fly' an "intensional" statement of some sort of "folk-scientific necessity"? But how would one measure whether different information is conveyed? Pelletier's answer is that we can test how the differing forms are used in tasks that involve default or nonmonotonic reasoning.

In chapter 5, James Hampton employs a prototype theory to investigate how generic concepts give rise to the sort of patterns one might expect from a "logical" account of the properties that are attributed to a prototype. In particular, Hampton is interested in the fact that when one group of people is asked to judge "how true" one generic statement is (e.g., 'Birds fly'), while another group is asked to judge "how true" a restricted version of that statement is (e.g., 'Dull-colored birds living in jungles fly'), the latter statement is, sometimes (or often), given a lower score than the former. The modifiers seem to have the effect of reducing the judged truth of these sort of generic statements. Not only do these results (and others that surround this effect) say important things about the "internal meaning" that people assign to generic statements on the basis of their combining concepts, but it also should say something about the "external meaning" in terms of the truth conditions of generics.

Susan Gelman approaches her study of generics in the language of children in chapter 6 with the observation that children acquire generic concepts with ease, despite being given ambiguous and insufficient evidence. Although this seems of a piece with general issues that arise in word learning—itself an instance of the classic problem of induction—there are many further peculiarities in the case of generics. Children start using generics between two and three years of age, and Gelman's view is that this capacity to understand them suggests that the input of particular types of generics might be the way that children's concepts and knowledge are shaped. The picture is that children seem to possess "kind" concepts beforehand that permit the early acquisition of generic noun phrases. Gelman sketches a number of ways in which one could investigate this hypothesis, by looking at the strength and scope of these effects. She also poses the interesting question of whether the types of generics being encountered by a child merely focus the child's attention in particular ways, or whether generics are capable of leading children to construct different kinds of concepts than they might otherwise do.

Mass Terms

After the introductory essay on the logical, linguistic, and philosophical considerations relevant to mass terms, three chapters explore relevant aspects of the cognitive importance of mass terms. In chapter 8, Cliff Goddard follows earlier work of his and of Anna Wierzbicka's (Wierzbicka 1988, 1996; Goddard and Wierzbicka 2002) in which they argue that there are numerous, subtly different, subclasses of mass nouns. These different subclasses presuppose, they claim, different "conceptualizations" of the nature of a mass. Chapter 8 concentrates on concrete mass nouns in English, arguing that the formal linguistic properties of mass nouns are systematically correlated with their conceptual content and that this conceptual content can be clearly identified using the tools of Wierzbicka's and Goddard's Natural Semantic Metalanguage (NSM) system. Goddard considers seven different subclasses of concrete mass terms and uses the NSM system to explicate the differing "unitizer" constructions that are relevant to the different subclasses. Some of the cognitive issues raised here involve the extent to which there is some unity to the notion of mass stuff and whether there is any necessary similarity in the relevant part of the mental lives of speakers of different languages, since it is clear that languages differ in the types of words that they each treat as mass.

Edward Wisniewski also asks about the conceptual basis of a mass/count distinction in chapter 9. He lays out three different attitudes that might be taken about this. One is that the grammatical distinction is semantically opaque and unprincipled, so children just learn that some nouns are count while others are mass, and they do not have any understanding of *why* there is such a syntactic distinction. A second viewpoint is that the syntactic distinction is deeply based on a conceptual difference: when speakers use count nouns to refer to things, they implicitly have something in mind, common to all count nouns, that they are trying to communicate, and when they use a mass noun, they have something in mind, common to all mass nouns, that they are trying to communicate. Wisniewski argues against both these views and instead supports the position that the mass-count distinction is to a very large degree conceptually based, but that there are exceptions and that some of these exceptions can be seen as caused by competing communicative functions of language.

In chapter 10, Fei Xu takes on the cognition of mass terms by considering the acquisition in children of *sortal terms*. Sortal terms are an important subclass of the count terms (non-mass terms), namely, those that provide criteria for individuation and identity. In a way, these terms are all the things that mass terms are sometimes said not to be, and for this reason it is illuminating to see the ways that acquisition of such terms differs from or matches that of mass terms. Non-sortal terms, such as adjectives like 'red' and mass terms like 'oatmeal', cannot be counted unless there is some sort of sortal or classifier associated with them: we can count the number of red balls in a room but not the number of "(the?) red" in a room. We can count bowls of oatmeal on the table but not "the oatmeal" on the table. The research reported here investigates how representations of sortal concepts develop in infancy and how learning count terms plays a causal role in constructing these concepts. Xu suggests that the very early-learned vocabulary is continuous with the later-learned vocabulary, despite differences other researchers have noted, such as the slow growth of earlier learned words in contrast with later-learned worrds. However, Xu contends, this is more likely due to early processing difficulties, rather than due to fundamentally different processes and representations in the earlier stages of acquistion.

REFERENCES

Bach, Emmon. Natural language metaphysics. In Ruth Barcan Marcus, G.J.W. Dorn, and Paul Weingartner, editors, *Logic, Methodology, and Philosophy of Science, VII*, pages 573–595. North Holland, Amsterdam, 1986.

Cohen, Ariel. More than bare existence: An implicature of existential bare plurals. *Journal of Semantics*, 23:389–400, 2005.

Goddard, Cliff, and Anna Wierzbicka. *Meaning and Universal Grammar: Theory and Empirical Findings*. John Benjamins, Philadelphia, 2002.

Greenberg, Yael. *Manifestations of Genericity*. Routledge, New York, 2003.

Lawler, John. Generic to a fault. In *Proceedings of the Eighth Meeting of the Chicago Linguistics Society (CLS-8)*, pages 247–258. Chicago Linguistics Society, Chicago, 1972.

Lawler, John. Tracking the generic toad. In *Proceedings of the Ninth Meeting of the Chicago Linguistics Society (CLS-9)*, pages 320–331. Chicago Linguistics Society, Chicago, 1973a.

Lawler, John. *Studies in English Generics*. PhD thesis, University of Michigan, 1973b.

Laycock, Henry. Some questions of ontology. *Philosophical Review*, 83:3–42, 1972.

Laycock, Henry. Theories of matter. In Francis Jeffry Pelletier, editor, *Mass Terms: Some Philosophical Problems*, pages 89–120. Reidel, Dordrecht, 1979.

Laycock, Henry. *Words without Objects*. Oxford University Press, Oxford, 2006.

Prasada, Sandeep. Acquiring generic knowledge. *Trends in Cognitive Sciences*, 4:66–72, 2000.

Prasada, Sandeep, and Elaine Dillingham. Principled and statistical connections in common sense conception. *Cognition*, 99:73–112, 2006.

Wierzbicka, Anna. Oats and wheat: Mass nouns, iconicity, and human categorization. In Anna Wierzbicka, editor, *The Semantics of Grammar*, pages 499–560. John Benjamins, Amsterdam, 1988.

Wierzbicka, Anna. *Semantics: Primes and Universals*. Oxford University Press, Oxford, 1996.

PART I

GENERICS

1

Generics: A Philosophical Introduction

FRANCIS JEFFRY PELLETIER

As remarked in the introduction to this volume, there are two different phenomena that have been comprehended under the title "generics": (a) reference to kinds and (b) some statements of generality.[1] Reference to kinds is a feature of (some) noun phrases (NP),[2] since it is (some) NPs that can

---

[1]. The issues in this overview are developed in much more detail in Krifka et al. (1995), from which the material here is adapted. The relation of generics to issues in nonmonotonic reasoning, mentioned but not developed later in this overview, is considered in Pelletier and Asher (1997). The two classical sources for modern treatments of generics are Lawler (1973) and Carlson (1980). A reasonably full bibliography of works on generics up to 1995 is in Carlson and Pelletier (1995).

[2]. Intuitively, and without invoking an entire linguistic theory, I picture some of the members of the lexicon of a language to be nouns (N), that these can be part of larger phrases called "common noun phrases" (CN), and that these in turn can be part of still larger phrases called "noun phrases." Intuitively, an N is a single word such as 'bear,' 'boy,' or 'bonnet' a CN is an N modified by such items as adjectives and relative clauses, such as 'brown bear' or 'boy who is tall' or 'white bonnet made of cotton' and a (full) NP is a CN that has been "determined" by a determiner or a quantifier, such as 'a big brown bear' or 'the boy who is tall' or 'some white bonnet that is made of cotton.' Two things might be noted: first, by this definition a plural added to a noun makes it a common noun, and second, this definition makes some pieces of language be simultaneously of two or all three of these categories. (E.g., although 'the boy' is an NP by this definition, the embedded

accomplish reference. Meanwhile, generic statements involve an entire sentence, since it is entire sentences, or at least entire clauses, that can make truth-claims. The two types can occur together because one uses generic statements to tell of a regularity that holds across individuals of a kind, and one way to state such a regularity is to predicate it directly of the kind. Thus, we might see a number of polar bears that are white, and none that are of any other color. One way to express one's feeling that this is "generically true" of polar bears is to attribute the property WHITENESS to the kind, *Ursus maritimus*, and say '*The polar bear is a white animal*'. We can see already one of the many philosophical puzzles arising: since species and genera are abstract objects, they cannot be white; only physical objects can be white. So, how is it that we can predicate whiteness of the kind, the polar bear? And yet another philosophical puzzle is in the background here, since we know that it could be that some polar bears have a genetic anomaly that makes them be brown rather than white. Yet, it could remain true that polar bears are white. How can generic statements be true, that is, report a feature of reality in this manner, while acknowledging the existence of exceptions?

Reference to Kinds

Reference to kinds occurs when an NP refers directly to a (abstract?) kind. Sentences that employ this sort of NP will attribute a property directly to the kind, and only indirectly, if at all, to members of that kind. How can we tell when that is happening? Consider the sentences

(1) a. The dodo is extinct.
    b. Shockley invented the transistor.

Here we are assured that the predicates '*is extinct*' and '*invent*' apply directly to a kind, because they are not applicable to individuals at all—no individual can be extinct, only species or kinds; no individual item is invented, only the type or kind. Thus, the subject of (1a), '*The dodo*,' and the direct object of (1b), '*the transistor*,' must refer to kinds. Once we have assured ourselves that there really is at least some direct reference to kinds, it becomes natural to see many other types of sentences as involving the same mechanism:

(2) a. *The potato* was first cultivated in South America.

'boy' is both a CN since it is CNs that get "determined" and become NPs, and also are N because that is the lexical entry that could have been modified by an adjective or relative clause. A "bare plural" such as 'bears' in the sentence '*Bears hibernate*,' is simultaneously a CN and an NP. A "bare singular" such as the mass term 'water' in '*Water is wet*' is simultaneously an N, a CN, and an NP.) When assigning semantic values to items of language, it is of course crucial that the values be of the sort that is relevant to the linguistic category to which the term belongs, so one should expect that the semantic value of these sorts of terms will change depending on which category is being discussed.

b. *Potatoes* were introduced into Ireland by the end of the seventeenth century.
   c. The Irish economy became dependent upon *the potato*.

all involve reference to the kind *Solanum tuberosum*. (Although there was probably some first-cultivated potato, and some first potato that came to Ireland, these seem not to be what the sentences in (2) are saying). We see that not only can definite NPs as in (2a) refer to kinds, but also that bare plural NPs as in (2b) can do so, too. And this reference is not restricted to the subject position of sentences, as (1b) and (2c) show. We should note also that mass terms can be kind referring, as in

(3)   *Gold* is a precious metal.

As well, we can use indefinite and quantified NPs to accomplish a special type of kind reference, that which is usually called "taxonomic kind reference," as in

(4)   a. The World Wildlife Organization decided to declare *a large cat* to be endangered.
      b. The World Wildlife Organization decided to declare *several large cats* to be endangered.
      c. *Three metals*—titanium, platinum, and iridium—moved up sharply on the commodities market.

One of the most interesting questions that reference to kinds brings forward is the issue of what the relation is between "ordinary individuals" and generic NPs. A standard answer is that the relationship is one of "exemplification": individual people exemplify the kind MANKIND, and a certain collection of individual houses exemplify the kind ARTS AND CRAFTS HOUSE. (The exemplification relation is normally left as a primitive concept.) One might wonder whether there are any other relations, and investigating the variety of generic NPs might throw some light on this.

It has been noted that it is not possible to form kind-referring NPs with just *any* noun. Contrast (5) with (6):

(5)   a. The German shepherd is a faithful dog.
      b. The Coke bottle has a narrow neck.

(6)   a. ?The German fly is a lazy insect.
      b. ?The green bottle has a narrow neck.

Basically, the noun (N) or common noun phrase (CN) must somehow be "semantically connected with" a *well-established kind* before one can make it a determined NP that designates a kind. But, of course, by constructing an appropriate story as background (e.g., a story that describes how medical science has discovered the way that green bottles protect medications indefinitely), one can thereby establish green bottles as a kind, and legitimately say

(7)   The green bottle has saved countless children's lives.

Does this suggest that kinds are created (and destroyed) by our use of language? Does it mean that an individual or a society can promote an NP to be kind referring? What implications are there in this observation for ontology, for relativism, and for realism versus antirealism? How do these sorts of considerations fit into the project of natural language metaphysics? A study of kind-reference, from both the theoretical and psychological points of view, could throw some light onto this philosophically important question.

As we have seen, reference to kinds is not simple, even in the case of direct reference to kinds that I have been discussing. But matters are even more complex in the case of "indirect reference" to kinds. There are very many different sorts of this indirect reference to kinds, and it is not at all transparent as to what causes some sentence—which is directly about an ordinary individual (in senses to be specified)—to indirectly refer to a kind. Is it a feature of the verb phrase? Or maybe culture? The differences are quite difficult to understand, but here is a classification, following Krifka et al. (1995).

The first type of indirect reference occurs when reference to a single, specific object generates reference to a kind. One intuitively thinks that this should not be possible, since a reference in a sentence to a specific object—which makes an individual predication relevant to that object—should result in a claim that is particular to that one object alone. Yet there are at least two different ways this sort of reference to an individual can be indirect reference to a kind.

(8) Representative Object Interpretation
   a. In Alaska we filmed *the grizzly*.
   b. Look children: *this* is the reticulated giraffe.
   c. Quiet! !—*The lion* is roaming about!

(9) Avant-garde Interpretation.
   a. *Man* set foot on the Moon in 1969.
   b. *Man* learned to solve cubic equations in the thirteenth century.

In the sentences of (8), only one object need be relevant, and yet this is sufficient to generate a truth about the kind. In (9), the claims are true of the kind MAN because of the actions of some first particular instance of that kind performing the action in question. The name "avant-garde" suggests that this type of kind reference is somehow essentially temporal, although it is difficult to make this precise, since not everything that was done for the first time by some person can generate this indirect reference to a kind, as we will see shortly.

There are also references to kinds that happen (apparently) because of the sort of *property* being predicated. I here mention four different types. In these cases, we manage to refer to the kind even though the property in question is the sort that applies just to individual members of the kind. The four types of reference I consider differ in the number, or distribution, of members of the kind to which the property applies:

(10) Characterizing Property Interpretation.
   a. *The potato* contains vitamin C.
   b. *Scandinavians* are blond.

(11) Distinguishing Property Interpretaton
   a. *The Dutchman* is a good sailor.
   b. *Italians* are good skiers.

(12) Collective Property Interpretation.
   a. *The German consumer* bought 11,000 BMWs last year.
   b. *Linguists* have more than 8,000 books in print.

(13) Average Property Interpretation
   a. *The American family* contains 2.1 children.
   b. *German teenagers* watch four hours of TV daily.

In the characterizing property interpretation (which I will consider in more detail further below), the property applies to the "typical" member of the kind—although perhaps not to all members. In the distinguishing property case, we manage to refer to the kind, the DUTCHMAN or the ITALIAN, even though the property being considered applies only to some (perhaps very small) subset of Dutchmen or Italians. The idea is something like this: even though most Italians don't ski at all, those who do ski distinguish themselves by being very good at it (by international standards). The collective property interpretation projects a property to the kind by means of a summation of the property's holding of all instances of that kind, while the average property interpretation projects a property to the kind from an averaging of related properties of members of the kind.

In psychological experimentation, then, it is important to determine what sort of kind reference is being employed, and to not inadvertently mix different types. In the converse direction, it would be very helpful for semanticists to have some sort of understanding derived from the psychology of speakers that explained what it is about their understanding or perception of the different NPs that gives rise to these different interpretations.

These seven different types of kind reference can be further distinguished. Note first that direct kind reference can involve a taxonomy, and hence employ a plural:

(14) *The/Some dinosaurs* are extinct

which means that all (or some) of the species of dinosaurs are extinct. With the characterizing interpretation, an indefinite NP can be used with a meaning equivalent to the definite NP: (10a) means the same as (15).

(15) *A potato* contains vitamin C.

Here, the expectation is that all or most of the typical instances of potatoes will manifest the property of containing vitamin C. In this way, characterizing predication differs from the distinguishing, collective, and average interpretations, where there is no such expectation. Distinguishing interpretations are different from both characterizing and average interpretations. Note that if one uses the indefinite in the distinguishing sentences, the result is *false*, even though the bare plurals are true.

(16) a. *An Italian* is a good skier / *Italians* are good skiers.
b. *A Frenchman* eats horsemeat / *Frenchmen* eat horsemeat.

This suggests that, in this sort of context, an indefinite means something like "A typical or randomly chosen X will (probably) Y" (which is a characterizing meaning), whereas the distinguishing interpretation itself plays upon some presumed-known discriminating feature. (11a) means (something like)

(17) a. *The Dutch* are known to have good sailors.
b. *The Dutch* distinguish themselves from comparable nations by having good sailors.

The avant-garde interpretations have two further unusual features. First, the property in question has to be "important" for the kind. Thus, although the two properties mentioned in (9) clearly apply to mankind, the properties in are quite dubious:

(18) a. ?*Man* broad-jumped more than 8.8 m in 1968.
b. ??*Man* ate 37 hot dogs in 12 minutes in 2005.

What is it about these properties and kinds that make such predications infelicitous? It seems not to be a matter of what holds in reality, since all the examples I've used in the avant-garde discussion are based on factual events. But then, how does this play into psychological accounts of the ways people understand generic reference? Is there anything that can be gleaned from experimentation in this area?

The other unusual feature concerns the nature of the kind in question. Neil Armstrong, the avant-garde object that makes (9a) true, is not only a person but also an American and a mammal. Yet both sentences in (19) are improper, for some reason:

(19) a. ??*The American* set foot on the Moon in 1969.
b. ??*The Mammal* set foot on the Moon in 1969.

An interesting finding from psychology could be to show why '*Man*,' but not '*The American*' or '*The mammal*,' designates a kind appropriate for this sort of predication. Once again, such a finding would be of enormous interest to researchers in philosophical and linguistic semantics.

Generic Statements

The other notion of genericity concerns sentences that neither report specific or isolated facts, nor quantify over such facts, but rather express a kind of general property. They report a regularity that summarizes *groups* of particular episodes or events or facts or states of affairs. Much of our commonsense knowledge of the world is expressed by these generic sentences, and this is what makes them especially interesting to epistemologists and psychologists who are interested in understanding how people encode information about the world—as well as to artificial intelligence researchers interested in constructing artificial agents who would be capable of operating in the "real world." Consider the following:

(20) a. Potatoes contain vitamin C.
b. The lion has a mane.
c. Machines are made from metal.
d. Fred drinks wine with dinner.

Not only are these distinct from individual or particular predications that might be made about some specific potato, lion, or machine or about some particular dinner of Fred's, but also they differ from explicit quantificational sentences such as

(21) a. *Each* potato contains vitamin C.
b. *All* lions have manes.
c. Fred *always* drinks wine with dinner.
d. *Some* potatoes are purple.
e. *Many* psychologists are clinicians.

And, as I remarked above, the two forms of genericity can occur together:

(22) a. *The potato* is highly digestible.
b. *Potatoes* are served whole or mashed as a cooked vegetable.
c. *The lion* has a mane.
d. *The Ivy-League humanities professor* wears a tweed jacket.

One of the most notable features of generic sentences is that they are "exception tolerant": Fred might omit wine from a few of his meals, some lions do not have manes, some potatoes are indigestible, and so on. In such circumstances, the sentences in (20) would be true while the corresponding ones in (22) would be false. It is this feature that piques the interest of many logically oriented linguists, philosophers, and artificial intelligence researchers.

But how many exceptions can a generic statement tolerate and still be true? Consider the following "squish" of examples:

(23) a. Snakes are reptiles.
b. Telephone books are thick.
c. Guppies give live birth.
d. Lions have manes.
e. Italians are good skiers.
f. Frenchmen eat horsemeat.
g. Unicorns have one horn.

If there were just one counterexample to (23a), we would say it was false, but (23b) is true despite the large number of communities with thin phonebooks.

In fact, only female guppies give birth at all, and just the impregnated ones at that; and only adult male lions have manes, and even some of them have had accidents that caused them to lose their mane. It therefore seems that (23c) and (23d) are true of somewhat less than half the relevant population. As I remarked before, the fact is that only some few Italians ski; furthermore, only some few Frenchmen eat horsemeat. But the Italians who ski competitively are very good, and the institution of eating horsemeat is deeply a part of French folklore. Finally, *no* unicorn has one horn.

Even a "vague" quantifier would fail. Consider the vague quantifiers '*Generally*' or '*In a significant number of cases.*' The following generic statements, without any such quantifier, are false, yet each would be true if quantified by one of these vague quantifiers:

(24) a. Leukemia patients are children.
b. Seeds do not germinate.
c. Books are paperbacks.
d. Prime numbers are odd.
e. Crocodiles die before they are two weeks old.
f. Bees are female.

There furthermore seems to be an "intensional" aspect to this sort of genericity. Consider the following generic statements:

(25) a. This machine crushes oranges.
b. Mail for Antarctica goes in this box.
c. Members of this club help one another in emergencies.
d. Children born in Rainbow Lake, Alberta, are left-handed.
e. Pandas have exactly three legs.

The sentence (25a) can be true despite the machine's being destroyed just as it emerges from the production line, never to actually crush any oranges. (25b) can be true even if there happens never to have been any mail destined for Antarctica, and similarly (25c) can be true even if there never has been a relevant emergency. The statements are true because of a *purpose* of the machine, or an *agreement* as to where the mail is to be put, or *preparedness* to help in emergencies. Conversely, even if all the children born in Rainbow Lake happened to be left-handed, that by itself would not make (25d) be true. For truth, we'd need to become convinced that there was something in the water of Rainbow Lake (or the like) that caused left-handedness, and it wasn't just a statistical accident. And again, it might turn out that, in the future, when there are only pandas left in the world, all in captivity, these very few pandas managed to all lose one of their legs because of an unfortunate series of accidents. Still, even though all the pandas had three legs, this would not make (25e) be true. Considerations such as these have suggested to some that generic sentences are akin to scientific laws: "accidental generalizations" do not make true generic sentences.

Related somehow both to this and to the issue surrounding (5) and (6) above (concerning the way that reference to kinds requires that the kinds be "semantically well established"), and equally difficult to explain, is that correct generic predication has to somehow be essential to the subject, unless the subject *directly* refers to a kind (the example is from Lawler 1973).

(26) Generic Predications.
    a. Definite NP reference
        (i) The madrigal is polyphonic ("essential" predication)
        (ii) The madrigal is popular ("accidental" predication)
    b. Bare Plural Reference
        (i) Madrigals are polyphonic ("essential" predication)
        (ii) Madrigals are popular ("accidental" predication)
    c. Indefinite NP
        (i) A madrigal is polyphonic.
        (ii) ??A madrigal is popular.[3]

This seems to show that there are different routes to take in the attempt to refer directly to kinds. As (26c-i) illustrates, one can make a generic characterization about a kind while using an indefinite NP, as long as the predicate is an essential or definitory of the subject. But if the property is merely an accident, as in (26c-ii), then—even though (26a-ii) and (26b-ii) are true statements—the claim no longer makes the same generic sense. (Unlike the definite NPs in (26a) and the bare plural ones in (26b), the indefinite version seems to make sense only when the predicate is a part of the "definition" of the subject. Since popularity is not such a property of madrigals, (26c-ii) comes out as infelicitous.)

It seems, then, that there is no number and no percentage that would tell whether or not a generic statement is true. So, what does? That is a question formal semanticists have been trying to answer. One therefore wonders whether any relevant information can be gleaned from the study of the situations in which people use generic sentences, or perhaps from the situations in which children learn to use these constructions.

Consider three generic sentences (words in italics indicate focal stress):

(27) a. Leopards usually attack monkeys *in trees*.
        'In cases where leopards are attacking monkeys, it is usually in trees.'
    b. Leopards usually attack *monkeys* in trees.
        'In cases where leopards are attacking something in trees, it is usually monkeys.'
    c. Leopards usually *attack* monkeys in trees.
        'In cases where leopards encounter monkeys in trees, they usually attack the monkeys.'

We see that different generic statements can be made by altering the stress of a sentence. The terminology used in the glosses brings forth the notion of "restricting cases," and it is employed in explaining the conditions under which the relevant events take place. It is not, for example, correct that leopards are

---

3. Of course, this is just fine were 'A madrigal' taken to refer to some specific individual madrigal, rather than to the kind.

usually attacking monkeys in trees. In fact, like most cats, leopards sleep more than half the time. Similar features come out when we consider the following:

(28) a. Tabby (usually) lands on her feet.
b. Marvin (normally) beats Sandy at ping-pong.
c. Bears with blue eyes are (normally) intelligent.
d. People who have a job are usually happy.

Tabby is *not* usually landing on her feet; Marvin is *not* normally beating Sandy at ping-pong. Instead, Tabby usually lands on her feet *in those cases where she is dropped*; Marvin normally wins *in those cases when they are playing ping-pong*. It is only in these classes of events that the main effect is being judged to usually or normally happen. In generics involving individuals (e.g., 28c), rather than events, we take the individuals to constitute the cases. This can lead to ambiguities, as in (28d), where on the individual reading we are talking about people who have a job and are saying that most of these people, or the typical ones, are happy people. On the other reading, we are talking about times during which a person has a job, and are saying that during most of these time periods the person is happy.

Krifka et al. (1995) develops a notation that seems adequate to capture these sorts of differences within the generic sentences:

(29) $\text{GEN}[x_1 \ldots x_i]$ $(\text{Restrictor}[x_1 \ldots x_i]; \exists[y_1 \ldots y_j]\text{Matrix}[\{x_1\} \ldots \{x_i\}; y_1 \ldots y_j]$

GEN here is an "unrestrictive quantifier" that binds all the x_k variables simultaneously.[4] (Although the examples I have been considering have only had one type of thing or event being quantified over, in theory there could be many.) In object-oriented generic cases, such as (28c), the variable being quantified by GEN ranges over objects, and the restrictor describes what objects are of interest; thus, (28c) would be represented as (30a) in this notation. If the cases being quantified were event structures, as in (28a), matters could be more complex, bringing into play more complexity in the variables and perhaps bringing into use the existentially quantified variables that reside in the matrix:

(30) a. $\text{GEN}[x]$ $(\text{Bear}(x) \& \text{Blue-eyed}(x)]$ $\text{Intelligent}(x)$.
b. $\text{GEN}[x]$ $(\text{Event}(x) \& \text{Dropping}(x) \& \text{Patient}(x,t); \exists y$ $(\text{Event}(y) \& \text{SubEvent}(y,x) \& \text{Culmination}(y,x) \& \text{Agent}(y,t) \& \text{Land-on-feet}(y,t)])$

Expression (30a) says that, generically, blue-eyed bears are intelligent. (30b) says that, generically, events that are droppings-of-Tabby contain a subevent whose culmination has Tabby landing on its feet. The specifics are not particularly important, for much could be altered and still retain the

---

4. The notion was made popular in Lewis (1975), to which the reader is directed for more details.

underlying intuition. What is needed, however, is some account of what it is to generically quantify, that is, an account of what GEN means.

Krifka et al. (1995) consider a number of alternatives and reject them all, leaving GEN an undefined notion. Here are some proposals that were rejected:

(31) a. A restriction to *relevant quantification*. In this proposal, the GEN is a universal quantifier, but the restrictor contains a "relevant predicate" that varies depending on the local context. For example, the generic sentence 'Whales give birth to live young' would become 'Every whale with property **R** gives birth to live young.' In this context, property **R** would be something like *'is female,' 'is adult,'* and so on. The problem with this approach is that it makes all generic statements true, since one can *always* find such an **R**. Consider 'Whales are sick': we choose **R** to be *'suffers from x, y, ...'* or maybe more simply *'is sick.'*

b. Assertions about *arbitrary objects*. The idea is to consider the existence of an object that has none of the properties that vary among the specific objects that exemplify the kind; for instance, an arbitrary whale would not be gray, or black, or white, or any other color that varies among whales. Nor would it have any particular size, and so on. The theory of abstract objects says that a generic claim amounts to asserting that such an abstract object would nonetheless have the property of giving live birth. Against this proposal one might object to the introduction of arbitrary objects into the realm of objects in the first place, and we might also note that the proposal seems unable to distinguish between accidental generalizations and legitimate claims of a generic nature.

c. Assertions about prototypes. In this proposal, generic statements are really claims about prototypes or about prototypical instances of kinds. Thus, 'Cats have tails' amounts to saying 'The prototypical cat has a tail.' To have this theory make any strong claims, one needs a robust theory of prototypes. Among other things, it would need to determine whether there is one or many prototypical cats. If the former, the generic statements are predications of this object; if the latter, then generic statements are universal quantifications over the set of these objects. Against the former version, note that both 'Ducks have colorful feathers' and 'Ducks lay whitish eggs' are true generic statements, and yet only male ducks have colorful feathers while only female ducks lay whitish eggs. Assuming that no duck, especially not a prototypical one, can be both male and female, it seems to follow that there must be more than one prototype of duck. But universal quantification now won't work, either, since it is not the case that all the prototypical ducks have colorful feathers, nor do they all lay whitish eggs. (Several chapters in this section presuppose the use of prototypes for the interpretation of these generics [e.g., chapters 3 and 5]. So, a task arising from this conference is to reassess the correctness of this critique of prototypes as the semantic value of generic sentences.)

d. Assertions about *stereotypes*. In this picture of generic quantification, stereotypes are *not* features of the world, but rather our conception or perception of the world. For example, 'Lions have manes' is seen as true, despite the fact that fewer than half of all actual lions have manes, because our perception/picture or stereotype of a lion is of something that has a mane. But there are various apparent shortcomings of this theory. Although 'Lions have manes' is true in this theory, 'Lions are male' is false despite the knowledge that the males are a superset of the ones with manes. And 'Lions are five-year-old males' is also false according to the theory, even though this is a subset of the ones with manes. Furthermore, Krifka et al. (1995) take the most telling objection to the stereotype theory to be that it makes the truth conditions for generics to be societal (or even personal) whims. 'Snakes are slimy' would be a true generic because of the stereotype. And even though we might admit that *in fact* snakes are not slimy, the existence of the stereotype would continue to make the generic statement be true.

e. Generics as *modal conditionals*. There is a clear similarity between generic statements such as 'Birds fly' and conditionals such as 'If x is a bird and is not abnormal, then x flies.' In turn, this conditional might plausibly be analyzed as 'In any of the most normal possible worlds, every bird flies.' Variations on this general theme have been pursued by researchers in the field of artificial intelligence, since it seems relevant for an account of intensionality and law-likeness. However, as Krifka et al. (1995) points out, several roadblocks lie in the path of this approach. Does the notion of *the* most normal possible world(s) make sense? Is it *really* more normal to have all birds fly? This can be done by killing off all the kiwis, ostriches, emus, penguins, and so on. Is that more normal? Or is it more normal to make them fly? To do this would be to either alter the structure of these birds drastically, or to change the laws of physics. Which is more normal?? And what about broken-winged birds, or fledglings? Is it really more normal either to eliminate them or to have them fly? If the most normal world had only bright-colored winged ducks, then there would be only male ducks and no female ducks. But then there would be no males, either. The details of this approach require much more work.[5]

---

5. One direction to this approach is taken in Pelletier and Asher (1997).

Perhaps some of the problems with one or another of these approaches can be removed by a careful reidentification of the underlying notions. Perhaps there are some other available interpretations of GEN. And finally, it is always possible that the whole approach under consideration—with its reliance on finding aspects of the world that constitute the "truth-makers" for generic statements—is a flawed approach to meaning, at least in this realm, and that generics are *not* to be evaluated as talking about "external reality," but rather about a "linguistic reality" or a "social reality." Many of the remarks in chapters 2 and 3 are relevant to this issue, as are the remarks from a child developmental perspective in chapter 6. As well, chapter 5 is directly aimed at defending a mentalistic picture of prototypes, and this, too, is directly relevant to the topic.

These are some of the sorts of results that would be of great interest to philosophers, linguistics, and semanticists generally.

REFERENCES

Carlson, Gregory. *Reference to Kinds in English*. Garland Press, New York, 1980.

Carlson, Gregory, and Francis Jeffry Pelletier, editors. *The Generic Book*. University of Chicago Press, Chicago, 1995.

Krifka, Manfred, Francis Jeffry Pelletier, Gregory Carlson, Alice ter Meulen, Gennaro Chierchia, and Godehard Link. Genericity: An introduction. In Gregory Carlson and Francis Jeffry Pelletier, editors, *The Generic Book*, pages 1–124. University of Chicago Press, Chicago, 1995.

Lawler, John. Studies in English Generics. PhD thesis, University of Michigan, 1973.

Lewis, David. Adverbs of quantification. In E. Keenan, editor, *Formal Semantics of Natural Language*, pages 3–15. Cambridge University Press, Cambridge, 1975.

Pelletier, Francis Jeffry, and Nicholas Asher. Generics and defaults. In J. van Benthem and A. ter Meulen, editors, *Handbook of Logic and Language*, pages 1125–1177. North Holland, Amsterdam, 1997.

2

Generics and Concepts

GREG CARLSON

In the experimentally oriented literature on concepts, one often runs across discussions that include language to the following effect:

> The concept 'dog' is characterized (in part) by 'has four legs.'
> 'Eats meat' is a feature of LION.
> 'Flies' is a part of the prototype of BIRD.
> ...

The language employed in such descriptions fairly regularly consists of one part (the concept) that is expressed as a noun or a short phrase based on a noun, and the other part that is a property, expressed as a predicate. (A relation is expressed between them, the nature of which may vary and that need not concern us immediately.) The point is that when these components are combined and dressed up a little to make them sound like grammatical English, the resulting sentences sound something like these:

1. a. Dogs have four legs.
   b. Lions eat meat.
   c. A bird flies.

Sentences like these express generalizations; they are not about particular events or characteristics, which might be exemplified by sentences such as the following:

2. a. Paul's dog is out in the yard.
   b. That lion is eating some meat I just gave it.
   c. A bird flew by.

These are not generic sentences. What is striking is that the conceptual literature tends not to use sentences like these in discussions (at whatever the level of formality intended in the discussion). Rather, the language employed to discuss concepts and their structures bears a striking resemblance to the language examined in the literature on generics and habituals (e.g., Krifka et al. 1995). That is, taking a very broad (and hazy) perspective, there appears to be a striking confluence of interests between the study of concepts in the psychological and cognitive science literature, and the study of certain types of sentences in the formal semantics literature. Despite these initial confluences, the relationship between the meanings of generic sentences and the study of conceptual structures remains unclear. While I cannot presently express any systematic ideas about what that relationship might be, this chapter explores this relationship a little further, gives some reason to think that further exploration of the relationship is warranted, and makes some tenuous suggestions about ways this exploration might be pursued.

The term *meaning* even as applied to natural language is notoriously difficult to define. However, I assume the normal machinery of what is currently known as *formal semantics* or *model theoretic semantics* in all its currently practiced varieties (which include dynamic approaches [Chierchia 1995; Groenendijk and Stokhof 1991; Kamp and Reyle 1993], situation semantics [Barwise and Perry 1983; Kratzer 1989], and the various offshoots of Montague grammar [Dowty et al. 1981; Partee 1974]). This body of literature focusing on natural language semantics, for all its variety, is theoretically fairly coherent and well understood, sharing fundamental assumptions and a kind of commitment to the agenda of truth-conditional semantics. It embodies a relatively clear view of what "meaning" (or at least a major component of it) might be, that is, that sentence meanings can be characterized in terms of truth conditions, which in turn provides the basic machinery for characterizing the denotational meanings of other constituent types. So while denotational meaning has a decently clear articulation to serve as a guide, putting meaning in terms of "concepts" seems at times little more than a relabeling.

1. "Concepts"

The term *concept* itself is also notoriously difficult to pin down; there is no single generally accepted intended sense to guide everyone in its application. There are, however, some things we can say. At the most general level, concepts are spoken of in two major ways, which I will call *objective* and *subjective*. The former are usually intended in philosophically oriented discussions of concepts (though not always), where it is truly, without abstraction, possible to talk about two different individuals apprehending, grasping, coming to understand (etc.) the same concept. It is "out there," independent of individual psyches (though quite possibly, then, "out there" is social, i.e., represented by psyches in a distributed fashion); it is what is represented when we speak of a mental representation *of* a concept. In the formal semantics literature, any identification of Fregean senses with concepts has this character.

On the other hand, the discussions of concepts in the psychological and experimental literature tend to assume that what is being studied directly in observing behavior are the concepts themselves. In this sense, two individuals cannot share the same concepts any more than they can share the same sense impressions or experiences. While we *can* talk of individuals sharing concepts (sense impressions, experiences), this is talk in abstraction; all we can really do is determine whether individuals A and B have similar concepts, and that's pretty much it—they can never be truly identical. (And then, of course, one can easily take the representational view, where we call the objective things "out there" "concepts," and then apply that selfsame term to the representations of those concepts themselves. This is where things tend to become a little confusing, for it tends to mix up what the exact object of study really is.)

For present purposes, I use *concept* to refer to subjective properties of individual psyches, because that seems to be the most dominant assumption within the experimental literature I wish to connect with here. This choice in no way endorses any view that *concept* should be confined to this sense alone, or that there is anything suspect about discussion of objective concepts. It is a matter of the perspective one wishes to take.

But in making this practical terminological choice, one gains only a little in terms of the clarity with which the term is defined. One way to illustrate the range of choice remaining is to operationalize the notion, that is, to ask "What do you want concepts to do for you?" "What do you want these to give some account of?" Depending upon one's answer, one gets very different ideas of what a psychological concept might be. At the most basic level, it appears that everyone wants concepts to give an account of categorization. But, for example, does one also wish for the structure of concepts to account for vagueness of category boundaries as well? Does one wish for concepts to give an account of semantic entailment? Here, for instance, are some other kinds of things that Ray Jackendoff (1989) explicitly asks psychological concepts to give some account of:

- our ability to understand language
- the calculation of consistency with other sources of knowledge
- the performance of inferences
- the formulation of responses
- translating structures from one language (or domain) to another

Prasada and Dillingham (2005) ask of concepts that they enter into explanations of the character of things in certain ways or, as Murphy (2002) puts it, that concepts express a theory of how things are (the "knowledge" approach to concepts). Wittgenstein (1952, ¶570) even more broadly suggests, "Concepts lead us to make investigations; are expressions of our interests, and direct our interests"; that is, concepts direct our activities in a global sense. Depending on how one answers these questions about what to expect from a theory in terms of human (and animal) behavior, one gets very different views of how "rich" a theory of concepts needs to be. I am going to leave this issue fairly open here and turn momentarily to generic sentences.

2. A Brief Semantics for Generic Sentences

Generic sentences take a wide variety of forms. Some of the most common include the following:

3. Cats have fur.
4. Mice squeak.
5. Policemen carry nightsticks.
6. The digital computer is found in many homes.
7. A crayon melts all too easily in the sunlight.

In each instance, the subject of the sentence is an expression that does not make direct reference to any particular individual of that class. Thus, 'cats,' 'the digital computer,' and 'a crayon' in these examples are understood as applying to or picking out no one particular individual (though they may easily be so understood, in other contexts).

Broadly speaking, there are two views on what the semantics of such sentences might be. The first is the quantificational approach. This assumes that a quantifier that has variable-binding properties sets conditions on the number or proportion of individuals that may satisfy the formula, in order to give an account of truth-conditions. Thus, (3) might be rendered as follows:

3'. $Qx[cat(x) \rightarrow have\text{-}fur(x)]$

Roughly, (3') schematically claims that there is some **Q** pattern of substitutions of individuals for the x that will make [cat(x) → have-fur(x)] true (e.g., in the "all" pattern, any substitution will be evaluated as true). (3') is imprecise in any number of ways, as it does not introduce intentionality and/or defeasibility, and I evade the precise nature of the connective (here rendered as "→"; for more details, see Krifka et al. 1995). But the essential point is that this analysis treats the subject as a predicate and not as a referring expression. Here, **Q** might be thought of as being an adverbial like "normally" or "usually" (i.e., 'cats normally or usually have fur' seems an approximate paraphrase). The truth-conditions of such a formula depend crucially on substituting individuals as values of x (and, intuitively, only those individuals that are cats would have any bearing on the truth or falsity of the whole formula).

Another view is the one I have been espousing for some time. On this view, the subject noun phrase is treated not as a predicate but rather as a fully referential expression which denotes a kind of thing.[1] Thus, the noun phrase 'cats' will refer to the kind CATS (represented here as c), and predication will be attributed without mediating quantification. Thus, (3) might be represented by:

3″.   Have-fur(c)

The cost of approaching the semantics of generics making use of this strategy is that one needs to give some account as to why any predicate that can be meaningfully applied to a regular individual can also be applied meaningfully to a kind. This requires that there be some type-raising or type-shifting operators, whereas this correspondence follows naturally from the quantificational approach:

8. *Fluffy* has fur. *Cats* have fur.
9. *Mickey* squeaks. *Mice* squeak.

However, the quantificational approach does not give an adequate account of a number of other factors, which include the aspectual category of the resulting sentence, interactions with negation, and sources of intensionality (for one extended discussion, see Wilkinson 1991).

Rather, I'd like to turn to the question of how closely we can identify the meanings of these generic noun phrases (whether denotational or predicational) with the corresponding concepts. Certainly, if this were a reasonable hypothesis to pursue, then it would give us an excellent start on understanding why there seems such an affinity for phrasing property relations in the conceptual literature making use of generic sentences.

One preliminary observation about the type of language used to describe concepts and affiliated properties is that they all tend to be rather brief, whereas brevity is no more a feature of generic sentences in any language than of any other types of sentences. So not only do we talk about flying geese and meowing cats and leaf-bearing trees, but we also talk about:

10. *Unpainted kitchen appliances that are just beginning to rust* need to be replaced within a period of two to three months by a qualified kitchen professional in order to prevent any possibility of bacterial contamination.
11. *Friendly but slightly confused medical professionals without appropriate training who nonetheless have medical degrees from top-ranked teaching hospitals* have much to contribute to society beyond their incomplete medical expertise.

---

1. "Kind" is intended here more in the technical logical sense (Carlson 1977) than in a purely linguistic sense as explored in, for example, Wierzbicka (1996).

The principled point is that linguistic expressions can be unbounded in complexity—limited by the syntax of the language—with the consequence that, if these are mapped directly onto certain types of mental states that we are calling concepts, then we stand in need of a device (let us call it *conceptual combination*) that can produce the arbitrarily large number of corresponding brain states in order to provide these phrases with appropriate denotations. While many people find this a perfectly reasonable and even desirable consequence (e.g., Jackendoff 1990), the philosophical underpinnings of the truth-conditional semantic enterprise makes such an outcome highly undesirable (and, in fact, wrong). The very notion of truth is itself an indexical one, and it is not wholly dependent on aspects of human mental representations. For instance, to take an example from generic sentences, imagine that the vast majority of people have a belief that snakes are slimy. One can even imagine (contrafactually) that everyone believes this of snakes. But what determines whether they are in fact slimy is not how people categorize them or what people might believe about them, but rather the nature of snakes themselves (which are, in fact, not slimy). Moving from truth to reference, the extensive literature on proper names stemming from Kripke (1970) and the literature on natural kind terms stemming from Putnam (1975) expressly show the inadequacy of any theory of meaning for such things that depends upon people's mental states or abilities. The fundamentals of the theory of quantification and of modality likewise develop shortcomings if thought of as mentally interpreted constructs. Partee (1980) explicitly discusses whether linguistic meaning in general can be represented adequately by purely psychological notions. She shows, to put it briefly, that the space of meanings is simply too "big" to be represented by any brain or mind states (assuming that the mind/brain is finite in resources). While she couches her demonstration in terms of individuals, times, and possible worlds, it bears some emphasis that her conclusions do not depend on which particular parameters of evaluation one chooses, but will generalize to all sets of parameters that individuate meanings at least as finely as these. Further, it is well known that, if anything, these parameters divide up meaning too coarsely, and not too finely. Thus, her demonstration possibly understates rather than overstates the volume of the corresponding mathematical space necessary to define linguistic meanings.

Just because we cannot identify concepts with linguistic meanings does not mean that concepts must play no role whatsoever. In fact, I believe that the hazy confluences noted above between talk of concepts and generic sentences is not coincidental, and that the nature of concepts can play a serious role in determining how some generic sentences are understood. Maintaining an objectivist (denotative) semantics accounting for truth-conditions (and maybe more), the question becomes one of what corresponds in the mind to the meanings of phrases denoting kinds and phrases denoting their features. It is most plausible to think that kinds correspond to something in the mind, and these seem to be typically labeled concepts. In the next section I outline some phenomena associated with the semantics of generic sentences that might well be understood as connected to or influenced by conceptual structures.

3. Generic Sentences and Concepts

3.1. k- and t-Properties

Prasada and Dillingham (2005) explore in some depth the connection between kinds and their properties. Their research strongly suggests that people represent a principled connection between the type of thing something is and some of its properties. They call these kind or *k-properties*. However, other properties that the kind may have are not judged to call into account a principled sort of connection, but depend upon simple "statistical" cooccurrences alone (or primarily). They label these latter properties type or *t-properties*. For example, our judgment that dogs have four legs enters into certain types of explanation (e.g., this has four legs *because* it is a dog), and into normative judgments (if this is a dog, it *should* have four legs). On the other hand, our judgment that dogs are brown is an instance of a t-property, because this judgment seems to be based on sheer statistical regularity alone and does not enter into the explanatory or normative accounts. Here are a few other examples drawn from their article:

k-Properties ("Principled")	t-Properties ("Statistical")
airplanes have wings	barns are red
trains travel on tracks	cars have radios
cheetahs run fast	dogs wear collars
doctors diagnose ailments	golfers wear plaid pants

Now, it is not that properties ascribed to kinds are k-properties or t-properties once and for good. Rather, it depends upon what kind of thing they are being predicated of. For instance, redness said of barns appears to be a t-property, whereas redness said of blood would appear to be a k-property. Thus, the distinction between these types of properties is the consequence of predicating one thing of another, that is, of the meaning of a whole sentence. The point of the Prasada and Dillingham (2005) article is to substantiate this distinction experimentally.

3.2. Greenberg (2003)

Yael Greenberg has done some work on the formal semantics of generic sentences that appears to lead us to something that looks like it could well be the same distinction. She is interested in giving an account of an interesting and systematic difference in interpretation between English indefinite singulars as subjects of generic sentences, and bare plurals. For the most part, the matter of singular versus plural to the side, the two are pretty often substitutable for one another in a generic sentence, and they will seem to mean almost exactly the same thing, as the following illustrate:

12. A dog has four legs / A grizzly bear hibernates during the winter.
13. Dogs have four legs / Grizzly bears hibernate during the winter.

However, as has been known at least since the observations of Lawler (1973), there are many instances where the two differ, and the indefinite singular version simply seems strange or possibly ungrammatical:

14. a. Madrigals are popular / Rooms are square / ?Men are blond / Uncles like marshmallows
    b. #A madrigal is popular / #A room is square / #A man is blond / #An uncle likes marshmallows

Lawler originally characterized the relationship required by the indefinite singular as one where the property is "essential." But Greenberg revises this understanding, since essence is too strong, to one of requiring a "principled connection" for the properties. The contribution of her work is to characterize the notion of *principled connection* in a model-theoretic framework.

Part of Greenberg's case makes use of observations about *nonce* categories: categories often expressed by fairly complex noun phrases that carve out little-used or unusual categories. What, for instance, follows from a person being a French writer born in 1954, apart of whatever follows from being a French writer, and being born in 1954? Yet, for most such ad hoc categories, one could find fairly consistent properties associated with that group that would be judged to be simply accidental. So, suppose that, by sheer happenstance, French writers born that year by and large, and in contrast to those born in 1953, 1955, and so on, write quite technical papers. If we use the bare plural construction, even as unmotivated a connection as this sounds quite fine, whereas the indefinite singular examples (when interpreted generically) are seriously degraded:

15. a. French writers born in 1954 write very technical papers.
    b. #A French writer born in 1954 writes very technical papers.

Or, to exhibit one example from Greenberg, consider the following contrast:

16. a. Famous carpenters from Amherst give all their sons names ending with 't.'
    b. #A famous carpenter from Amherst gives all his sons names ending with 't'

Now, this does not mean all such ad hoc categories lead to strangeness in the indefinite singular. If one perceives a principled connection between the kind and its predicated property, it becomes fine. Consider the phrase 'bananas that have been sat on by a rhinoceros,' a category with very low frequency of occurrence (a Google string search reveals no matches whatsoever at the moment). If we say

17. Bananas that have been sat on by a rhinoceros are flat,

one can easily understand the principled connection: the sitting causes the flatness of the otherwise-shaped banana. Thus, we predict that the indefinite singular expression of the corresponding sentence will be acceptable:

18. A banana that has been sat on by a rhinoceros is flat.

So it is not a matter of whether the category is ad hoc, but rather whether a principled connection exists.

If we return to the Prasada and Dillingham (2005) materials and substitute indefinite singulars for the subjects of their proposed k- versus t-properties, we intuitively seem to find that the t-property examples are judged to be less acceptable. Consider the following sampling drawn from their categories of natural, artifactual, and social kinds:

k-Properties
    A car has four wheels.
    A train travels on tracks.
    A cheetah runs fast.
    A lemon is sour.
    An artist is creative.
    A gymnast is flexible.

t-Properties
    A barn is red.
    A shower cap is transparent.
    A pigeon sits on statues.
    A rock is jagged.
    A biker has tattoos.
    A Hindu lives in India.

It would appear that Prasada and Dillingham, and Greenberg are talking about much the same thing.

3.3. *Property Stability*

Lyn Frazier has pointed out to me that t-properties appear to be less "stable" than judged k-properties over time. That is, while most generic properties are amenable to change, the k-properties are, on the whole, more persistent than the t-properties. If, for instance, bears ceased hibernating, they would still be bears, but one would have the distinct impression that there must have been a big change (whether in bears or the environment) to account for this. On the other hand, if one noted that bears no longer performed in circuses, no corresponding big change would be assumed because one was not assuming this to be a highly stable property (being a t-property).

Now, it is also a fact that generic sentences, while commonly expressed in the simple present tense in English, can be also expressed in the past (or in the future, or without tense, for that matter). However, if one uses a generic in the simple past tense (or future) that could have been easily used in the present as well, consider the intuitive effect:

19.
   a. Dogs ate meat.
   b. Teachers assigned homework.
   c. Airplanes had wings.

The force of the present observation depends on the extent to which the intuitions about the examples are shared; in this respect, it would be very nice to have quantified data to present as well. In practice, though, the clearer the intuitions, the less inclined one is to regard the need for quantification necessary, and the intuitions expressed here definitely fall at the less clear end of the scale. A typical reaction to (19a) might be to understand the utterer as implying that dogs no longer eat meat (or that at least it's an open question). This effect seems to vanish if one adds an explicit past time adverbial (e.g., 'Even before they became domesticated, dogs ate meat').[2] The examples in (19) would also be true in the present. In (20), in contrast, we presume the present tense to be inappropriate, perhaps because there are no longer dinosaurs, ancient Romans, or Pony Express riders.

20.
   a. Dinosaurs laid eggs.
   b. Ancient Romans wore togas.
   c. Pony Express riders were expert horsemen.

One senses some difference between the examples of (19) and (20). In examining (19) in contrast to (20), we see two possible effects. First, we can find the examples of (19) a little odd or unacceptable, as if some presumption is being made that does not hold. Second, we could, quite possibly, also accord them a nongeneric interpretation as a means of trying to avoid this oddness, because in nongeneric sentences a similar effect does not usually occur (e.g., if it's true that 'John fed the dog' at some past time, it does not seem to preclude John feeding the dog in the present). Understanding them this way would yield interpretations similar to saying '*some* dogs ate meat (on a certain occasion),' or '*some* airplanes had wings.'

Frazier and Clifton (n.d.) have run a preliminary study that applies this observation to the study of k- and t-properties. They compared the k-properties list of Prasada and Dillingham and their t-properties with sentences that have been put into the past tense. Since t-properties are more amenable to change, the reasoning was that it would be simpler to imagine a circumstance where they are no longer true but once were. However, k-properties, because of their relative stability, would make it more difficult to naturally

---

2. My thanks to an anonymous reviewer for the point, and the example itself.

interpret them generically in the past tense, making a nongeneric interpretation more likely. In the study, this is what was found. The full results await further work. Thus, it appears that the interpretations of certain types of sentences of natural language can be understood as affected by a distinction arising from work on the nature of concepts.

3.4. "Suitability"

Ariel Cohen (2005) uses the term *suitability* to refer to the use of existential uses of bare plurals versus existentially quantified noun phrases in the attributed property of generic sentences. Cohen notes that in certain instances substitution of a quantified existential for a bare plural will yield a slight degradation of judgment, whereas in other instances there is no similar degradation. For example, (21a) seems more acceptable than the quantified instance of (21b) (substitution of 'some wheels' for 'wheels'):

21. a. This tractor has wheels.
    b. (?) This tractor has some wheels.

In contrast, there seems no such similar contrast between the examples found in (22):

22. a. These ambulances have dents in them.
    b. These ambulances have some dents in them.

The judgments are subtle but, I believe, stable enough. (Cohen offers no quantified data, and I find the intuitions elusive at times.) Cohen explains the contrast this way: suppose the tractor in question in (21) had only two wheels. This is not a "suitable" arrangement of wheels for the proper functioning of a tractor. Under these circumstances, the speaker would be more apt to phrase the sentence as (21b) than as (21a). Put roughly, how the wheels on a tractor are arranged, and how many there are, can affect the tractor's ability to function fully in the way intended. On the other hand, there is no such similar contrast in (22), since dents in ambulances, however arranged, however many there are, do not contribute to the functioning of the ambulance in the way intended (though, of course, huge dents might inhibit their functioning).

Lyn Frazier again has noted to me that if one divides up the predicates according to whether they are t-properties or k-properties, the "suitability" implicature is or is not triggered. The properties that appear to have the characteristics of k-properties accorded them by Prasada and Dillingham (2005) are the ones that appear to trigger the implicature, whereas the t-properties do not. Frazier and Clifton (n.d.) again are working on quantitative means of checking on this, with encouraging results. The hypothesis to be evaluated is whether t-properties yield fewer preferences for bare plurals over "some" noun phrases than do k-properties, which ought to show a distinct preference for the bare plurals.

3.5. A Long-Standing Puzzle

In Carlson (1977), I puzzle about the following example. If you look at cars, you will clearly see that

    23. Cars have tires.

This is true in most people's experience for 99+% of all cars observed. It is also true to that exact level of consistency that the tires observed were all black (white-wall tires are black, too, by the way, but they are "out" these days and so we'll ignore them). So while it seems to be true that cars do indeed have tires, and it is also true to nearly 100% consistency that the tires are black, there is something nonetheless strange about the following sentence:

    24. (?) Cars have black tires.

For those who get the intuition (and some don't), the sense is that the blackness of the tires is somehow irrelevant. But hold on:

    25. Cars have air-filled tires.

This seems quite fine by comparison, yet why might 'black' seem irrelevant when 'air-filled' is not? After all, almost exactly as many tires on cars are air-filled as are black. These distinctions cannot be accounted for, obviously, by some appeal to number, frequency, probability, or proportion. However, if we examine these examples from the standpoint of suitability implicatures, and k- and t-properties, we do see a distinction. The blackness of tires does not further the utility of the vehicle, and it does not make it function better in the way intended than if it had green tires; on the other hand, the tires being air-filled does enhance the suitability of the car to function appropriately—let the air out and you soon see why. Further, if one thinks about whether 'black' as applied to tires is a k- or t-property, one is most likely to conclude that, despite its consistency as a property of tires, it is a t-property; on the other hand, 'air-filled' as a property of tires would in most peoples' judgment be likely to come out as a k-property.

4. Generic Reference and Concepts, Perhaps

The preceding section discussed some linguistic phenomena that seemed to correlate, in some cases rather convincingly, with the k- and t-property distinction discussed and motivated in Prasada and Dillingham's (2005) work. In this section I wish to discuss a couple of potential applications of conceptual structures for noun phrase reference. Recall that the general hypothesis is that generic noun phrases are taken to refer to kinds of things, and that these phrases can be of unlimited complexity; hence, one can refer to an unlimited number of kinds of things. This appears to be true for generics that make use of bare plural expressions in English.

However, as has been pointed out for some time, the alternative definite generic expressions of English are limited. Krifka et al. (1995) express this limitation as a requirement that English definite generics refer only to "well-established kinds." That is, there is some presupposed familiarity, or something of that character. This suggestion is not more precisely defined. As an illustration, consider the following contrasts:

26. a. *The bottle* has a narrow neck.
    b. *The Coke bottle* has a narrow neck.
    c. ??*The green bottle* has a narrow neck.
    d. OK *Green bottles* / *Coke bottles* have narrow necks.

27. a. *The Indian elephant* has smallish ears and is easily trained.
    b. ??*The friendly elephant* is easily trained.

The examples in (26) are originally due to Barbara Partee. They illustrate a distinction not only between 'the bottle' but also between 'Coke bottles' and 'green bottles.' Coke bottles are a familiar type of bottle (or presumed to be), whereas green bottles are not so regarded.

The basic problem here is, however, that at least at the time the examples were first introduced, Coke bottles were (light) green in color, and thus everyone's experience with Coke bottles would have been a subset of experiences with green bottles. Thus, green bottles ought to have been more, and not less, familiar. Note that this distinction disappears when bare plurals are used (26d). Or, suppose Indian elephants are all friendly elephants (and some others are, too); in this case, one's experience with Indian elephants would have been no more frequent than one's experience with friendly elephants, yet even on this understanding the distinction remains. Though I've been phrasing things here in terms of familiarity (of experience), it seems clear that to be "well established" involves something different, or at least far more.

Let us return to the question left in abeyance above regarding how concepts (as construed here) might participate in the semantics of language. We have already set aside the hypothesis that they serve as the denotata of linguistic expressions in general, but this does not preclude the possibility that they could serve as the denotations of parts of language, or be parts of meanings (just not the whole story). Greg Murphy (2002) explicitly argues what is implicit in most discussions of concepts, that concepts are the denotations (the meanings of) the lexical items of a language. Certainly this is plausible; discussion of lexical semantics is at the root of most linguistic work that invokes concepts as explanatory elements, and in the psychological work on concepts, lexical semantics is by far the most common vehicle used to express a concept (as when one writes, 'the concept GREEN'). More specifically, the kinds of lexical items used tend to be almost exclusively drawn from the nouns, verbs, adjectives, and occasionally the prepositions of English and similar languages.

Let us assume, only for the moment, that concepts as meanings are limited to the lexical items of the language. On this supposition, 'green,' 'dog,' and 'jump' will denote concepts (and hence there must be the concepts GREEN, DOG, and JUMP), but such syntactically complex expressions as 'greener than your mother-in-law's kitchen wall,' 'hungry or thirsty dog,' and' jump right over the fence' do not denote concepts. This leaves open the possibility that there are such concepts as HUNGRY- OR- THIRSTY- DOG, but there need not be in order for the corresponding phrase to have a meaning, unlike the monomorphemic lexical items. The consequence is, then, that complex phrases would not take concepts as their meanings, if concepts as meanings are limited only to meanings of lexical items. Thus, we would not (contrary to most common views of concepts) expect 'thirsty dog' to designate a concept, if each of its component parts names one.

Put more to the point, the hypothesis under consideration is that concepts do form meanings for certain types of expressions of natural language but not for others. They might fail to serve as denotations for two reasons: one would be that the semantics of the language does not map certain category meanings, or it might be that there are no such corresponding concepts. Above I briefly suggested that these be limited to only the lexical items of the language, but this might be at once too strong, and/or too weak.

Let us consider the possibility that the constraint on the denotation of English definite singular generics is that the whole noun phrase denotes a kind if and only if the common noun it is based on itself denotes a concept (or, rather, it is presupposed that there be a concept denoted by the common noun). Thus 'the lion' will be an acceptable generic noun phrase because the 'lion' part will itself denote the concept LION. But the generic noun phrase 'the friendly elephant' will not be acceptable (as seems correct) because there is no concept denoted by 'friendly elephant.'

As the examples above indicate, the definite generics are not limited to monomorphemic expressions, because 'the Coke bottle' and 'the Indian elephant' are acceptable. The former is a compound noun and thus would qualify as a lexical item of the language. However, 'Indian elephant' has the structure of an adjective-noun combination, and it is clear that not all phrases of this structure form acceptable definite generics (as the unacceptability of 'the green bottle' and 'the friendly elephant' would indicate). And, it does not seem entirely clear that anything that has a compound noun structure necessarily forms an acceptable definite generic:

28. a. ??The actress scandal
    b. ??The alcohol syndrome
    c. ??The shift boss
    d. ??The orange peel garbage bag
    (Compare:
    Actress scandals are more common than they used to be.
    A typical alcohol syndrome involves tremors.
    Meet Bob, he's my shift boss.
    You categorize your garbage well: is this your orange peel garbage bag?)

It is entirely possible to create contexts in which these do not sound so bad, but it appears to require establishing a presupposition of the appropriate sort (e.g., that orange peel garbage bags have a specific design that requires some ingenuity to create for the purpose). And, there are intensionalized definite descriptions that can further make for false-positive judgments. So, if it is not structure that determines whether something is an acceptable definite generic of English, then it must have something to do with the meanings expressed. In this, one has the common intuition that there is a perceived difference between 'green bottle' and 'Coke bottle,' or 'Indian elephant' and 'friendly elephant.' It goes something like this: an Indian elephant is one of a recognized variety of elephants, and their properties are not simply those of being an elephant, and being from India, but something more (e.g., disposition, size of ears, etc., etc.)—it's a (sub)species. In this sense, 'Indian elephant' differs from 'friendly elephant' because a friendly elephant is no more than an elephant that is friendly, and that's it. 'Indian elephant' also differs from a phrase like 'Indian tree' in that an Indian tree is simply something that is a tree and from India, and that's it, so the phrase 'the Indian tree' ought not be a good definite generic, which seems correct.

It is reasonable to think that the constraint might be one on requiring that the denotations be "natural kinds," but one must then accept artifacts ('the digital computer,' 'the fountain pen') and socially created things ('the sonnet,' 'the jump shot' [in basketball]) as natural kinds as well. I'm suggesting, for further thought, that perhaps the term *concept* should be countenanced as the critical ingredient in place of *natural kind*. In any event, there are clearly further constraints that neither gives proper account of. It is, for instance, difficult to naturally apply these to many human categories ('??The lawyer is indispensible in our system of justice,' '??The violin player...'), and, as Vendler observed (1971), terms that are "too general" do not seem to sit well as definite generics:[3]

29. a. *The parabola* is easily plotted.
    b. ??*The curve* is easily plotted.

5. One Last Phenomenon

The last matter I wish to consider does not concern generic sentences specifically, but it is related to genericity by virtue of the involvement of determinerless nominals, one of the main ways of expressing kind-reference across languages. The phenomenon in question is that of noun incorporation, in which (typically) a direct object determinerless noun is

---

3. An anonymous reviewer notes the following do not seem bad:
   i. The curve entered architectural design in the Baroque era.
   ii. The curve proved a stumbling block for early geometers.

I find (i) quite good, contra Vendler's claims; (ii) is not as good for me.

morphologically joined to a verb, becoming a part of it (e.g., Baker 1988; for two overview articles, see Gerdts 1998; van Geenhoven 2002, and references therein). One example of this is given below, in the Chukchi language:

30. a. Enan qaa-t qErir-ninet
        3sg.Erg reindeer.Abs.pl seek-3sg.s/3.pl.o
     b. Etlon qaa-rer-g'e
        3sg.Abs reindeer-seek-3sg.s
        'He is looking for reindeer.' (Spencer 1995)

In (30a), the determinerless noun *qaa-t* is in "normal" direct object position, whereas in (30b) the nominal, stripped of its plural marking, is morphologically made into a part of the verb. Both sentences mean about the same thing, though with some subtle differences (beyond the issue of plurality) we'll come back to in a moment. It could be rendered only roughly in English (which does not have a direct equivalent) as 'He is seeking reindeer' versus 'He is reindeer-seeking.'

In the substantial literature on incorporation, it is often noted that the process is not fully general (though at times it can be). In particular, it is often noted that the noun–verb combination must describe what is variously characterized as a "familiar," "habitual," or "generic" activity, or that it describes an activity that is "nameworthy"; it is occasionally described as designating a "unitary concept." These are, of course, notional characterizations that, it appears, point to very much the same phenomenon. For example, in Chukchi, if one uses the incorporating form of 'reindeer + kill,' it does not mean simply to kill a reindeer (as the unincorporated form does), but to kill a reindeer as a ritual part in the preparation of a meal (Dunn 1999). That is, in the Chukchi culture, it is a "nameworthy" or "habitual" sort of activity.

Nameworthiness is also a means of dealing with lexical gaps. I will illustrate the point using Norwegian, as studied in great detail in Borthen (2003). While Norwegian does not have incorporation proper, it does allow determinerless singular count nouns in object positions (English is much more restricted in this). The semantics of the determinerless nominals, though morphologically not incorporated, appear to be identical to the semantics of incorporated nominals in other languages (e.g., Chung and Ladusaw 2004; Farkas and de Swart 2004; van Geenhoven 1998). They are "semantically incorporated" forms.

Borthen notes that not all combinations of predicates and bare nominals are acceptable. While, for instance, the examples in (31) are natural and acceptable, those in (32) are not:

31. a. Han anbefalte rullestol.
        He recommended wheelchair.
        'He recommended a wheelchair.'
     b. Jeg kan lese bok, jeg.
        I can read book I.
        'As for me, I can read a book.'

c. Han eier bil.
      He owns car.
      'He owns a car.'
30. a. *Per slo jente.
       Peter hit girl.
    b. *Hun vasket sykkel ren.
       She washed bicycle clean.
    c. *Ulven drepte okse.
       Wolf-the killed bull.

Borthen goes to great lengths to rule out a wide variety of possible accounts, and in the end provides an explanation for this particular class of constructions, which she calls the "conventional situation type" construction, along the following lines: The predicate and the noun together must denote a conventional situation type, and the construction can be reasonably seen as a multiword lexical entry. A conventional situation type is "a property, state, or activity that occurs frequently or standardly in a given contextual frame, and has particular importance or relevance in this frame as a recurring property" (p. 153).[4]

I believe that this same sort of characterization is, in other rubrics, more broadly noted in many studies of the nature of noun incorporation in many languages. Note that Borthen's sense that these could be regarded as multiword lexical entries is made more plausible in incorporating languages, where the noun/verb combination is in fact of the category of a lexical item, namely, a verb.

We have been considering the possibility that the lexical items of a language, at least, may denote or be associated in some grammatically significant way, with concepts. If we think of concepts as primarily classifications of sense experiences we might have, and that words of a language may be used to express or encode many of those concepts, we might take Borthen's characterizations and think in terms of the role a concept might play. Consider the possibility that the requirement for a verb–noun combination to be acceptable is that it presupposes that there is a corresponding concept. So, for instance, car owning and book reading would be such potential concepts, but bull killing and bicycle washing would not be. With such a provision, we could provide an account of such lexical gaps as those noted above, and in many other languages for incorporation and incorporation-like constructions. Of course, this is merely the rankest of speculation without there being some independent means of determining if one or the other is or is not a concept, and at the moment I must leave this up to the ingenuity of psychologists. But the gross outline of a view of

---

4. The analysis is, of course, much more nuanced than presented here. One would wonder, for instance, based on summary here, why "wheelchair-recommending" comes out as a "conventional situation type." See Borthen (2003) for a consistent analysis.

concepts that emerges from suggestions like this and those above is one where concepts are not, in general, available as the denotations of all types of natural language expressions. Instead, they are "spotty," can be idiosyncratic, and are to a great extent tied up with the nature and structure of a language and a culture.

In closing this section, we need to recognize that one major issue has been systematically skirted in the above discussions. We took the point of view that concepts were primarily psychological objects, and hence fundamentally unshared. However, the discussions above rely upon there being a means of sharing them, or at least aligning them, within a community of individuals speaking the same language. While I believe there are ways of accomplishing this that make critical use of the fact that people speak a common language, it remains, of necessity, an agenda for another venue.

6. Conclusion

Concepts, at least as studied by psychologists, have not found a natural place in the study of natural language semantics in the formal semantics tradition as practiced in the past thirty or so years. This does not point to any shortcomings in semantic theory, I emphasize, for there are excellent intellectual as well as practical reasons as to why this should be. However, it does leave us with something of a disconnect between what is going on in the mind—and I take concepts to be major features of it—and the way we talk about things.

This has been an attempt to make suggestions about how we might close that gap somewhat. We do not examine truth-conditional phenomena for evidence of concepts in semantics. Rather, we consider how truth-conditional phenomena are expressed in a language, and the restrictions and constraints on that, to uncover our clues. We also require a clearer articulation of what "concepts" in fact are. The extent to which work on natural language semantics can be put together with work on culture and work on the mind, is the extent to which we begin to have a chance of understanding what might be responsible for the phenomena discussed in this chapter.

ACKNOWLEDGMENTS

I thank two anonymous reviewers for substantial improvements in this chapter, and the panel discussion at the conference on Generics and Mass Terms, which stimulated many of the ideas here.

REFERENCES

Baker, M. (1988). *Incorporation*. Chicago: University of Chicago Press.
Barwise, J., and J. Perry. (1983). *Situations and Attitudes*. Cambridge, MA: MIT Press.

Borthen, K. (2003). Norwegian Bare Singulars. Norwegian University of Science and Technology, Trondheim, doctoral dissertation.
Carlson, G. (1977). Reference to Kinds in English. University of Massachusetts Amherst, doctoral dissertation.
Chierchia, G. (1995). *Dynamics of Meaning*. Chicago: University of Chicago Press.
Chung, S., and W. Ladusaw. (2004). *Restriction and Saturation*. Cambridge, MA: MIT Press.
Cohen, A. (2005). More than bare existence: An implicature of existential bare plurals. *Journal of Semantics* 23, 389–400.
Dowty, D., R. Wall, and S. Peters. (1981). *An Introduction to Montague Semantics*. Dordrecht: D. Reidel.
Dunn, M. (1999). A Grammar of Chukchi. Australian National University, doctoral dissertation.
Farkas, D., and H. de Swart. (2004). *The Semantics of Incorporation: From Syntax to Discourse Transparency*. Stanford, CA: Center for the Study of Language and Information.
Frazier and Clifton. (n.d.). A Role for k-Properties in Language Comprehension? University of Massachusetts/Amherst manuscript.
Gerdts, D. (1998). Incorporation. In A. Spencer and A. Zwicky, eds., *The Handbook of Morphology*. Oxford: Blackwell, 84–100.
Greenberg, Y. (2003). *Manifestations of Genericity*. New York: Routledge.
Groenendijk, J., and M. Stokhof. (1991). Dynamic predicate logic. *Linguistics and Philosophy* 14, 39–100.
Jackendoff, R. (1989). What is a concept, that a person may grasp it? *Mind and Language* 4, 68–102.
Jackendoff, R. (1990). *Semantic Structures*. Cambridge, MA: MIT Press
Kamp, H., and E. Reyle. (1993). *From Discourse to Logic*. Dordrecht: Kluwer.
Kratzer, A. (1989). An investigation of the lumps of thought. *Linguistics and Philosophy* 12, 607–653.
Krifka, M., F.J. Pelletier, G. Carlson, A. ter Meulen, G. Link, and G. Chierchia. (1995) Genericity: An introduction. In G. Carlson and F. J. Pelletier, eds., *The Generic Book*. Chicago: University of Chicago, 1–124.
Kripke, S. (1970). *Naming and Necessity*. Cambridge, MA: Harvard University Press.
Lawler, J. (1973). *Studies in English Generics*. University of Michigan Papers in Linguistics 1:1. Ann Arbor: University of Michigan Press.
Murphy, G. (2002). *The Big Book of Concepts*. Cambridge, MA: MIT Press.
Partee, B. (1974). *Montague Grammar*. New York: Academic Press.
Partee, B. (1980). Semantics—mathematics or psychology? In R. Bäuerle, U. Egli, and A. von Stechow, eds., *Semantics from Different Points of View*. Berlin: Springer, 1–14.
Prasada, S., and E. Dillingham. (2005). Principled and statistical connections in common sense conception. *Cognition* 99, 73–112.
Putnam, H. (1975). The meaning of "meaning." In K. Gunderson, ed., *Language, Mind, and Knowledge*. Minneapolis: University of Minnesota Press, 131–193.
Spencer, Andrew. (1995). Incorporation in Chukchi. *Language* 71, 439–489.
van Geenhoven, V. (1998). *Semantic Incorporation and Indefinite Descriptions: Semantic and Syntactic Aspects of Noun Incorporation in West Greenlandic*. Stanford, CA: Center for the Study of Language and Information.

van Geenhoven, V. (2002). Noun incorporation. *GLOT International* 8, 261–271.
Vendler, Z. (1971). Singular terms. In D. Steinberg and L. Jakobovitz, eds., *Semantics: An Interdisciplinary Reader*. Cambridge: Cambridge University Press, 115–133.
Wierzbicka, A. (1996). *Semantics: Primes and Universals*. New York: Oxford University Press.
Wilkinson, K. (1991). Studies in the Semantics of Generic Noun Phrases. University of Massachusetts Amherst, doctoral dissertation.
Wittgenstein, L. (1952). *Philosophical Investigations*. New York: Macmillan.

3

Conceptual Representation and Some Forms of Genericity

SANDEEP PRASADA

Despite the fact that we encounter only a very limited number of instances of any given kind, we acquire generic knowledge of the sort expressed by sentences such as (1).

(1) Dogs are four-legged.

How is generic knowledge of this sort acquired? A reasonable answer to this question requires, at least, a characterization of (i) the nature of the knowledge acquired and (ii) the conceptual mechanisms needed in order to represent knowledge of this sort. This chapter reviews some recent research that has sought to provide these prerequisites. It also discusses the nature of the conceptual mechanisms uncovered by this research and what they reveal about some forms of genericity and their linguistic expression.

Some Types of Mental Mechanisms

Before considering the question of how generic knowledge is represented, it is worth reflecting on the nature of the mechanism that allows us to think and talk about kinds and instances of kinds. It is a remarkable fact about human cognition that even though we encounter only particular entities, we naturally think and speak of those particular entities as instances of kinds.

For example, we may point to a particular entity and say something like

(2)   That is a dog.

In doing so, we identify the particular entity we encountered as being the same as indefinitely many other entities with respect to the kind of thing it is (e.g., a dog). Not all mental mechanisms function in this manner. For example, indexical mechanisms, such as the demonstrative *that*, as well as mechanisms of visual attention (e.g., object files and FINSTs [Feigenson and Carey, 2005; Kahneman et al., 1992; Pylyshyn, 1989; Scholl, 2001]), do not function in this manner. They provide the means for directing attention to a single entity (or what is represented as a single entity). They do not provide the means for thinking of indefinitely many entities as being the same in any respect. Thus, two uses of *that* do not require that the things that are referred to be the same in any respect, much less that they be the same with respect to the kind of thing they are. As such, indexical mechanisms do not provide the means for thinking of a particular entity as an instance of a kind (Prasada and Dillingham, 2009).

It is also important to note that though it is natural for us to think and talk about particular entities as instances of kinds, it is by no means necessary for intelligent behavior that an organism be able to represent particular entities as instances of kinds. Simply representing particular entities with their particular properties and responding to new particular entities on the basis of their similarity to previously encountered entities can allow an organism to behave in intelligent ways, without ever representing anything as an instance of a kind. This is the guiding idea behind exemplar theories of conceptual representation (e.g. Medin and Shaffer, 1978; Smith and Samuelson, 1997; Whittlesea, 1997). My point here is not to defend such a theory of conceptual representation.[1] It is only to highlight the fact that the mechanisms that allow us think and talk of particular entities as instances of kinds are formally distinct from the mental mechanisms exploited by exemplar theories to represent and store information about particular entities as such.

The Form of Basic Nominal Concepts

What, then, is the structure of the conceptual mechanism that allows us to think of a particular entity as an instance of a kind? It seems that a type-

---

1. A problem faced by theories of this sort is that different types of similarity are relevant to generalization in different domains, and thus domain-specific sources of constraint on representations of similarity are required for such theories to succeed. I agree with this point; however, the existence of domain-specific constraints does not require the representation of particular entities as instances of kinds. The domain-specific constraints may be on the form and/or content of the representation of particular entities as such without also requiring that the entities be represented as instances of a kind.

token mechanism such as the one in (3) is required (Prasada and Dillingham, 2009). Such a mechanism may be used to generate indefinitely many new representations, as in (4), each of which provides the means for thinking about a different instance of the same kind K.

(3) K_i
(4) $K_1 \; K_2 \; K_3 \ldots$

The representations in (4) are the mechanisms by which we think about specific *instances of kinds*. These representations can be linked to indexical and perceptual representations accounting for the fact that we can direct our attention to and perceive those entities that we think of as instances of kinds. On the other hand, (3) is the mechanism that allows us to think about a *kind* and thus cannot be directly linked to perceptual representations. We can think and talk about kinds but cannot perceive them.

The mechanism in (3) also provides the basis for a number of other fundamental properties of how we think about kinds. For example, it naturally captures the fact that we think that the identity of a kind does not depend on how many or which instances of the kind exist at any given time (e.g., we do not think that the kind DOG changes with the birth or death of each dog). The mechanism in (3), which provides the means for thinking about the kind, is capable of generating indefinitely many representations for thinking about instances of the kind (4); however, (3) clearly does not change in any way, no matter how many such representations are generated or by whether or not the things thought about via the mechanisms in (4) continue to exist. As such, (3) can ground the notion that kind identity does not change as instances of a kind come into and go out of existence.[2]

The structure of (3) also provides the basis for distinguishing kinds in a manner relevant to the count–mass noun distinction in language. Thus, one may distinguish between kinds for which we do and do not understand any two instances of the kind K_a, K_b to be quantitatively equivalent by virtue of being the kinds of things they are. In the former case, an instance of a kind can function as a unit of quantification, whereas this is not possible in the latter case. Thus, the conceptual basis for the count—mass noun distinction may be captured in terms of a constraint stated over the representations generated by (3).

Clearly, much more would have to be done to fully develop these suggestions. My purpose here is simply to point out that the mechanism in (3) affords the possibility of capturing these fundamental facts about the

---

2. Though the kind itself remains the same, it does not follow that all of the properties of the kind remain the same. Most obviously in the current context, the kind has different instances and, possibly, different numbers of instances in its extension at different times. Other properties associated with the kind may also change as a function of the changing composition of a kind; nevertheless, we think and speak of the kind as remaining the same. The mechanisms in (3) and (4) show how this may be possible.

ways in which we think about kinds and instances of kinds. As such, it has some motivation independently of the uses to which it will be put in developing an account of the mechanisms needed for representing generic knowledge of the sort expressed in (1).

The Need for Elaborating This Mechanism

The mechanisms in (3) and (4) allow us to think about things as instances of kinds. We are not limited, however, to simply thinking of particular entities as instances of kinds. We also think of them as having various properties (e.g., being brown, having four legs, or being hungry). Furthermore, some recent evidence from a neuroscience experiment using event-related potentials suggests that sentences that characterize kinds (e.g., 'Bananas are yellow') are processed differently than sentences that characterize instances of kinds (e.g., 'This banana is yellow') (Prasada et al., 2008). This raises the question of how we represent the relation between the kind of thing something is and other properties it possesses.

Prasada and Dillingham (2006) addressed this question from within the perspective of an explanation-based approach to conceptual representation. The guiding idea of the explanation-based approach is that concepts do not merely provide the means for sorting things into categories, but also provide the means for understanding what things are and why they have the properties they do. As such, concepts are understood to possess explanatory structure. Consequently, much research within this approach has sought to discover the modes of explanation embodied in our commonsense conceptions of things. This research has found that our conceptual systems exploit causal-essential, teleological, and intentional modes of explanation (e.g., Carey, 1985; Gelman, 1990; Gelman and Wellman, 1991; German and Johnson, 2002; German et al., 2004; Gergely et al., 1995; Gopnik and Melzoff, 1997; Inagaki and Hatano, 2002; Johnson and Solomon, 1997; Keil, 1989, 1994; Kelemen, 1999; Kuhlmeier et al., 2004; Leslie, 1994; Lombrozo and Carey, 2006; Opfer and Gelman, 2001; Springer and Keil, 1991, among many others).

Prasada and Dillingham (2006) proposed that our conceptual systems also contain a formal mode of explanation wherein certain properties of instances of a kind are explained by their being the kinds of things they are. For example, a dog is understood to have four legs because it is a dog, but it is not understood to be brown because it is a dog.[3] This suggests that our conceptual systems must represent two types of connections between the kind of thing something is and properties of instances of that kind: one that

---

3. For a discussion of how the formal mode of explanation is related to other modes of explanation, see Prasada and Dillingham (2006, 2009).

licenses *formal explanations* (explanations by reference to the kind of thing something is), and one that does not.

The connections that license formal explanations were dubbed *principled connections*, and the properties that bear a principled connection to a kind, *k-properties* (e.g., four-leggedness is a k-property with respect to the kind DOG). Connections that do not license formal explanations were dubbed *factual connections,*, and the properties that bear a factual connection to a kind, *t-properties* (e.g., being brown is a t-property of the kind DOG). Statistical connections are a type of factual connection (e.g., there is a weak statistical connection between being a dog and being brown).[4]

Distinguishing Principled and Factual Connections

In addition to licensing formal explanations, principled connections were hypothesized to license normative expectations concerning the presence of k-properties (e.g., we think dogs, by virtue of being dogs, *should* have four legs). Finally, principled connections license the expectation that instances of a kind will generally possess their k-properties. In sum, principled connections were hypothesized to have an explanatory aspect, a normative aspect, and a statistical aspect.

Statistical connections, even strong statistical connections, on the other hand, do not license formal explanations or normative expectations concerning t-properties. Thus, even though barns are typically red, we cannot explain the redness of a barn by citing the fact that it is a barn. Likewise, we do not think that barns, by virtue of being barns, should be red, or that there is something wrong with nonred barns.

Evidence for the Existence of Principled Connections

Prasada and Dillingham (2006) showed that it is possible to identify a set of kinds and properties for which participants paraphrase bare plural sentences such as (1) as meaning either (1a) or (1b). This suggests that participants understand these properties to be true of instances of a kind because they are the kinds of things they are, as in (1a), and that these properties are understood to generally be true of instances of the kinds in question, as in (1b). As such, the connection between these kinds and properties was shown to display two key characteristics of principled connections.

---

4. The "k" in k-properties indicates that these properties are determined by the kind of thing something is. The "t" in t-properties indicates that these are properties that are associated with a type to the extent that tokens of the type possess the properties. Note also that designation of a property as a k- or t-property is always relative to a kind. For example, red is a k-property of strawberries but a t-property of flowers.

(1) Dogs are four-legged.
   (1a) Dogs, by virtue of being the kinds of things they are, are four-legged.
   (1b) Dogs, in general, are four-legged.

(5) Barns are red.
   (5a) Barns, by virtue of being the kinds of things they are, are red.
   (5b) Barns, in general, are red.

For a different set of kinds and properties, such as the ones in (5), participants found only the general prevalence paraphrase (5b) to be appropriate, indicating merely a statistical connection between these kinds and properties.

In a second experiment, Prasada and Dillingham (2006) found that participants rated formal explanations of k-properties of (6) and (6a) to be much better than those of t-properties of (7) and (7a).

(6) Why does that (pointing to a dog) have four legs?
   (6a) Because it is a dog.
(7) Why is that (pointing to a barn) red?
   (7a) Because it is a barn.

Finally, in a third experiment, Prasada and Dillingham (2006) showed that participants have normative expectations concerning the presence of k- but not t-properties by showing that participants generally judge statements such as (8) to be true, but this is not the case for sentences such as (9).

(8) Dogs, by virtue of being dogs, should have four legs.
(9) Barns, by virtue of being barns, should be red.

Prasada and Dillingham's (2006) stimuli included items from a wide range of content domains (natural kinds, artifact kinds, social kinds). The same pattern of results was found in each domain, suggesting that the representation of principled and statistical connections is not limited to items of one or another kind, but may be represented for any kind of thing. Finally, for all of the experiments, virtually identical results were found for a subset of the stimuli that were matched on the average prevalence of the characterizing properties, showing that the results could not be due to differences in the prevalence of the k- and t-properties.

Representation of Principled Connections: The Aspect Hypothesis

Given that our conceptual systems represent principled connections between being a given kind of thing and certain properties, the question arises as to *how* principled connections are represented. Prasada and Dillingham (2009) hypothesized that representing a principled connection between a kind and a property requires representing the property as one aspect of being that kind of thing. They referred to this hypothesis as the *aspect hypothesis* (AH).[5]

The AH potentially helps explain the key characteristics of principled connections. Given that the existence of a whole presupposes the existence of its parts, the existence of a part (k-property) may be rendered intelligible by identifying the whole (kind of thing) of which it is a part. Thus, AH affords the possibility of exploiting a part–whole principle (PWP) to provide formal explanations of k-properties. If having four legs is represented as one aspect of being a dog, PWP allows us to explain the four-leggedness of any given dog by citing the fact that it is a dog.

The aspect hypothesis also potentially helps provide a natural explanation for our normative expectations concerning the presence of k-properties. If having four legs is understood to be one aspect of being a dog, then a dog *should* have four legs; otherwise, it would be incomplete and/or have something wrong with it. Thus, AH affords the possibility of exploiting a principle of perfection or completeness to ground normative expectations concerning the presence of k-properties.[6]

Finally, AH potentially explains the statistical aspect of principled connections. Insofar as having four legs is an aspect of being a dog, all dogs are expected to have four legs. However, having four legs is only one aspect of being a dog. Dogs, by virtue of being dogs, are also material entities and thus interact causally with their environments. In some circumstances, the causal interactions a dog happens to be subject to may prevent a k-property from developing or being present in a given instance. Because there is no reason to suppose that these special circumstances will generally prevail, dogs are expected to generally have four legs.

---

5. The term "aspect" was left undefined. The hypothesis, as well as the tests of the hypothesis conducted by Prasada and Dillingham (2009), requires only an intuitive interpretation of the term, meaning roughly *part* as in *not all of*. It is also important to note that AH does not, and is not meant to, provide a criterion for identifying k-properties. Instead, it proposes how principled connections between kinds and properties (identified in some other manner) are represented.

6. Prasada and Dillingham (2009) note that not all normative expectations are grounded in such a principle. For example, the notion that human beings should floss their teeth every day does not derive its normative force from such a principle. Daily flossing of teeth is not understood to be part of being a human being. Furthermore, one would not judge someone who does not do so to be incomplete in some way, or to have something wrong with them. In this case, the proposition derives its normative force from the fact that daily flossing is beneficial to someone. As such, it derives its normative force from a principle of beneficence. Though it clearly is good for instances to have their k-properties, the normativity that derives from k-properties being beneficial does not predict that instances that lack k-properties should be judged to be incomplete or have something wrong with them.

It is worth noting that t-properties of kinds receive no explanation within the formal mode of explanation. To explain t-properties, one must make use of a causal mode of explanation.

Evidence for the Aspect Hypothesis

In a series of experiments, Prasada and Dillingham (2009) provided evidence in support of AH. In a first experiment, they directly tested AH by asking participants to judge the truth of statements such as (10) and (11), which identified k- and t-properties to be aspects of being a given kind of thing. As predicted by AH, participants judged statements such as (10) to be true, but not statements such as (11).

(10) Having four legs is one aspect of being a dog.
(11) Being red is one aspect of being a barn.

A second experiment provided an indirect test of AH. If k-properties are represented as aspects of being a given kind of thing, it should be possible to characterize an arbitrary instance of the kind as possessing a k-property (12). On the other hand, it should not be possible to characterize an arbitrary instance as possessing a t-property (13), even if it is a prevalent t-property, because possessing the t-property is not understood to be an aspect of being that kind of thing. In line with the predictions of the AH, participants judged indefinite singular sentences containing a k-property (12) to be interpretable as characterizing either an arbitrary or a specific instance of a kind, whereas those containing a t-property (13) were only interpretable as characterizing a specific instance of a kind.

(12) A dog has four legs.
(13) A barn is red.

These results also provide experimental support for the long-noted observation that generic interpretations of indefinite singulars are possible when there is some type of special connection between a kind and a property (e.g., Burton-Roberts, 1977; Carlson and Pelletier, 1995; Cohen, 2001; Dahl, 1975; Lawler, 1973). Prasada and Dillingham's (2006, 2009) experiments help characterize the nature of this special connection and how it is represented in the mind. The results also provide experimental support for Greenberg's (2003) observation of the connection between "in virtue of" constructions and indefinite singular generics.

Further indirect evidence for AH came from an experiment that showed that participants judged informal definitions of kinds that included k-properties (14) to be better than those that included t-properties (15).[7]

---

7. It is important to note that the informal definitions we provide in response to the question 'what's an x?' do not provide necessary and sufficient conditions for being an x (Chomsky, 1996; McGilvray, 1999, 2005; Quine, 1987). As such, these are not definitions

(14) What's a dog? A dog is a four-legged animal.
(15) What's a barn? A barn is a red building.

These results along with those of the previous experiment also provide experimental support for Cohen's (2001) observation that indefinite singular sentences that can be interpreted generically allow formulation of definitions of their subject terms in terms of their predicate terms. It is important to note that the AH provides an explanation for why such a connection is found. Together, these three experiments provide strong evidence that representing a principled connection between a kind and a property requires representing the property as an aspect of being the kind of thing.

A fourth experiment showed that AH can account for the ability of principled connections to support formal explanations by providing independent evidence that representation of a part–whole relation licenses the possibility of providing formal explanations. The experiment showed that whereas participants do not think that one can explain the redness of a barn by reference to the fact it is a barn (because being red is not part of being a barn), it can be explained by reference to the fact that it is a red barn (because being red is indisputably part of being a red barn).

A final experiment provided evidence that the normative aspect of principled connections is underwritten by AH and reflects a principle of perfection or completeness by showing that instances of a kind that lack their k-properties are judged to be incomplete or to have something wrong with them.

In all of these experiments, the same pattern of results was found for a subset of stimuli that were matched on the prevalence of k- and t-properties. Furthermore, the same pattern of results was found in each content domain.

Conceptual Mechanisms Implicated by the Representation of Principled Connections

The experiments discussed above suggest (i) that our conceptual systems represent principled connections between being a given kind of thing and having certain properties, and (ii) that representing principled connections requires representing these properties as aspects of being a given kind of thing. Given that this is the case, and the assumption that a mechanism of the sort in (3) is needed in order to think of things as instances of kinds, a natural question that arises concerns how the mechanism in (3) must be

---

in a technical sense. What constraints there are on the form and content of such informal definitions is an empirical question.

elaborated so that it is able to represent principled connections between being a given kind of thing and having certain (k)-properties.

An alternative way of posing this question is to ask what kinds of conceptual mechanisms are implicated by the AH, that is, to represent the notion that a given property is one aspect of being a given kind of thing. Given that AH involves a part–whole relation of some sort between the k-property and the kind, one may be tempted to suggest that the conceptual representation of the k-property (e.g., YELLOW) is a constituent of the conceptual representation of the kind concept (e.g., CANARY) in the manner that YELLOW is a constituent of the phrasal concept YELLOW CANARY. Prasada and Dillingham (2009) argue against this suggestion. They point out that whereas tokens of phrasal types *must* have the properties denoted by their constituents, instances of a kind can lack their k-properties. Furthermore, following Fodor (1998), they point out that whereas it is clearly impossible to have the concept YELLOW CANARY without having both the concept YELLOW and the concept CANARY, one could certainly have the concept CANARY without having the concept YELLOW. These considerations make it clear that AH, and thus the representation of principled connections, does not entail that the *concept* for the k-property (YELLOW) is a part of the *concept* for the kind (CANARY).[8]

Prasada and Dillingham (2009) point out that AH requires that we represent one aspect of the things that are thought about via the concept CANARY as being yellow (a property that is thought of via the concept YELLOW). In other words, the part–whole relation is between the *things* that are thought about via the concepts YELLOW and CANARY rather than between the concepts YELLOW and CANARY (as is the case with the concept YELLOW and the phrasal concept YELLOW CANARY). The problem then becomes one of identifying the mechanism needed for representing a part–whole relation between the things thought about via two concepts without representing a part—whole relation between the concepts themselves.

Prasada and Dillingham (2009) propose a solution that is based upon the solution that linguists have developed for handling an analogous representational issue. In sentences such as (16), the nonidentical lexical items ('John,' 'him') may be interpreted as making reference to the same person. However, the nonidentical lexical items 'Bill' and 'him' cannot be interpreted as making reference to the same person.

(16) John$_i$ thinks Bill$_k$ likes him$_i$.

Given that this is the case, a mechanism is needed to capture whether or not nonidentical lexical items may be given identical interpretations. This is accomplished via the mechanism of co-indexation. Thus, 'John' and 'him'

---

8. For a more detailed argument for this point and other arguments regarding why lexical concepts such as CANARY are unlikely to be constituted of other concepts, see Fodor (1998).

may be interpreted as referring to the same person because they are co-indexed. The same mechanism can provide the means for implementing AH. When representing principled connections, however, what needs to be established is not the identity of the things thought about by two concepts (e.g., CANARY, YELLOW), but the identity of an aspect of the thing thought about via one concept (CANARY) and the property thought about via the other concept (YELLOW).

A mechanism of the sort shown in (17) provides the means for representing principled connections.

(17) $K_i \rightarrow a_1, a_2, \ldots P_{a_1}$

In this representation, the K represents the kind of thing something is, and the index i is a variable that can take on indefinitely many values, each representing a distinct instance of the same kind. The indexical elements a1, a2,... provide the mechanism by which we think about aspects of an instance of the kind K. The representation of a principled connection involves co-indexing one of these elements with the concept P that represents a k-property. For example, the principled connection between being a canary and being yellow is represented in the following manner (18):

(18) $\text{CANARY}_i \rightarrow a_1, a_2, \ldots \text{YELLOW}_{a_1}$

Prasada and Dillingham (2009) note that a desirable characteristic of (18) is that it clearly allows one to have the concept CANARY without also having the concept YELLOW. This highlights the fact that being yellow cannot be understood to be a *condition* for being thought of via the concept CANARY. It also allows for the possibility that an instance of a kind may lack a k-property. In contrast, one cannot have the complex concept YELLOW CANARY without having both the concept CANARY and the concept YELLOW, and being yellow is a condition for something to be thought of via the concept YELLOW CANARY. The concept YELLOW CANARY seems to require a mechanism such as the one in (19).

(19) $[\text{YELLOW} [\text{CANARY}_i \rightarrow a_1, a_2 \ldots]]$

The mechanism in (17) grounds the expectation that all instances of a kind will possess their k-properties by virtue of being the kinds of things they are, while allowing for the possibility that some specific instances of the kind may lack k-properties *for reasons other than their being the kinds of things they are*. What is ruled out by (17) is that an instance of the kind could lack being P by virtue of being a K.

Prasada and Dillingham (2009) suggest that a desirable characteristic of (17) is that the key elements that allow this mechanism to represent principled connections have independent motivation. They suggest that the elaboration of the representation of instances of kinds to include aspects is needed to capture the intuition that no matter how great the unity of a thing, things are always mentally divisible and thus can always be thought to have various parts or aspects. Furthermore, they suggest that the mechanism of co-indexation is needed to capture the intuition that certain kinds of properties (e.g., that denoted by YELLOW) seem to be existentially dependent. It is beyond the scope of this chapter to develop these suggestions. Instead, I'd like to consider how the mechanisms in (17) and (19) shed light on a number of phenomena concerning the representation of generic knowledge.

Four Ways of Thinking about Kinds

The mechanism in (17) captures the fundamental duality of our thinking about kinds of things (Macnamara and Reyes, 1994). A kind is at once thought to be a single thing and the indefinitely many instances that constitute it. This duality is captured by the form of the mechanism that consists of a fixed K that captures the oneness of a kind, and an index i that functions as a variable that may take on indefinitely many values that identify the indefinitely many instances of the kind that constitute the kind. This duality is also reflected in the ways in which we speak about kinds. In English, we can often talk about kinds of things via the use of a definite singular (20), highlighting the singularity of kinds, or a bare plural, highlighting the inherent multiplicity of kinds (21).

(20) The canary is yellow.
(21) Canaries are yellow.
(22) The Hindus consider the cow to be sacred.
(23) A canary is yellow.

We can also sometimes highlight the inherent duality of kinds via the use of definite plurals such as (22). Thus, it seems that the duality of our conception of kinds, which is captured in the form of (17), provides a natural explanation for why there are three perspectives from which we can think and talk about kinds: one that identifies a kind as a single thing, another as the indefinitely many instances that constitute the kind, and a third that identifies the unity in multiplicity of a kind. As we know, however, we can also think of kinds in terms of arbitrary instances (23). This raises two (related) questions. First, why is there a fourth way to think about kinds at all? Given the duality in our conception of kinds, it is clear why there can be three ways of thinking and talking about kinds, but why should there be a fourth way? Second, why is it that this fourth way is via characterization of an arbitrary instance of a kind?

The answers to these questions lie in the fact that the structure of (17) allows it to function either as a representation of a kind or as a representation of an arbitrary instance of a kind. In both cases, i is interpreted as lacking a specific value. In representing a kind, i is not interpreted as identifying a specific value because it may be identified with indefinitely many specific values and thus functions as a variable. On the other hand, in representing an arbitrary instance of the kind, i is not interpreted as identifying a specific value because it is incapable of having a specific value (i.e., an arbitrary instance simply lacks any property that is specific to any instance of the kind). Thus, we see that the mental structure that provides the means for thinking about a kind may also be used to think about an arbitrary instance of a kind. Furthermore, because an arbitrary instance lacks any information that is specific to any instance of a kind, it can be used to reason about what properties instances of a kind have by virtue of being instances of a kind.

We now have an answer to the questions concerning the characterization of kinds via the characterization of an arbitrary instance of a kind. There exists a fourth way of thinking about kinds because the mental structure that provides the means for thinking about kinds also provides the means for thinking about arbitrary instances of kinds, and the properties that can be attributed to an arbitrary instance of a kind are those that instances of the kind are expected to have by virtue of being the kinds of things they are.[9]

It is worth highlighting that the generality provided by arbitrary instances is not the consequence of quantification. Instead, it is the consequence of establishing that a connection exists between being a given kind of thing and having a given property that does not depend on the properties that distinguish one instance from another, but instead simply depends on being that kind of thing (Prasada, 2000).

Nature of an Arbitrary Instance

Given Pelletier and Asher's (1997) observation that "[t]he history of philosophy has not been kind to arbitrary objects" (p. 1139), it is important to make clear why the present account avoids the criticisms that have been leveled at the notion of an arbitrary object. In a nutshell, the various criticisms all point to the impossibility of such an object existing or its inability to represent instances of the kind in question.[10] For example, a (universally) arbitrary dog could not be black or brown or white or any specific color since all dogs are not of the same color and thus the arbitrary dog must be colorless; however, this is impossible because every dog must be of some color. On the other hand, if the arbitrary dog is supposed to represent dogs, it fails to do so because according to the representation dogs are colorless, which they are not.

---

9. (20)–(23) are linguistic reflections of four ways of thinking about kinds that follow from our conceptions of kinds and the manner in which they are represented. I am not claiming that there are only four linguistic devices for talking about kinds or that all languages use all of these linguistic devices—they don't (Chierchia, 1998; Krifka et al., 1995).

10. See Pelletier and Asher (1997) for a concise summary of the problems faced by theories of arbitrary objects.

What saves the present theory from criticisms of these sorts is that the mechanism in (17) is just that, a mechanism (of thought). It does not represent any instance of the kind that may exist in reality, imagination, or memory. Thus, questions concerning the color of the arbitrary dog simply do not come up. An arbitrary dog is not a dog, but rather a tool of thought that allows us to characterize what instances of a kind are like by virtue of their being the kinds of things they are. In the account developed here, an arbitrary instance of a kind does not and cannot exist except as a tool for thought. As a tool of thought, the mechanism can leave out of consideration properties that instances of a kind must have (e.g., color for dogs) without contradiction or misrepresentation. The fact that the mechanism does not, for example, make reference to the color of dogs does not mean that dogs do not have a color, only that when thinking and reasoning about dogs *qua* dogs, one need not consider the color of dogs. In sum, the present account avoids the criticisms aimed at theories of arbitrary objects because it does not countenance the existence of any such objects. Arbitrary objects are nothing but mechanisms of thought. That the appropriate way to think about arbitrary objects and their role in reasoning may be as tools of cognition has also been suggested by Tennant (1983).

Mechanisms for Thinking about Instances of Kinds and Members of Classes

The account developed above makes a formal distinction between two kinds of type-token mechanisms. The mechanism in (17) allows us to think of things as instances of kinds. In doing so, it allows us to think of indefinitely many numerically distinct entities as being the same with respect to the kind of thing they are. Furthermore, the mechanism provides the means for thinking of these things as possessing certain (k-) properties by virtue of being the kinds of things they are. On the other hand, the mechanism in (19) and its more general form (24) provide only the means for creating an equivalence class.

(24) [PROPERTY [KIND$_i$ → $a_1, a_2 \ldots$]]

It allows us to think of indefinitely many entities as fitting the description specified by the type (e.g., indefinitely many yellow canaries). The mechanism does not provide a basis for thinking that tokens of the type possess any properties by virtue of being tokens of that type (other than being tokens of that type and the logically necessary properties of being a token of that type). Thus, things thought about via the concept YELLOW CANARY must be yellow and must be canaries, and must have those properties that follow in virtue of being yellow as well as those that follow in virtue of being a canary; however, they are not understood to possess any other properties by virtue of being yellow canaries. Formally, this is because forming the phrasal concept does not introduce any aspects of the complex type that may be co-indexed with a putative k-property. Conceptually, it is because the mechanism in (19) does not provide the means for thinking about a new kind of existent—one that has certain properties because it is the kind of thing it is. Instead, (19) provides the means for thinking about the subset of the existents thought about via the concept CANARY that possess the property denoted by the concept YELLOW.

The form of (24) embodies the idea that there is no reason to assume that arbitrary subsets of a kind should share any properties over and above those that specify membership in the subset, much less that such properties, were they to exist, are true of members of the subset by virtue of being members of the subset. It is, of course, possible that some specific subsets share properties other than being members of that specific subset, but this would not follow on the basis of the *form* of the representation.

The difference in the formal structure of mechanisms (17) and (24) can provide a grounding for Prasada and Dillingham's (2006) observation that whereas concepts that allow us to think of things as instances of kinds (e.g., POLAR BEAR) have k-properties, concepts that merely provide the means for creating equivalence classes (e.g., WHITE BEAR) cannot have k-properties. For example, (25) sounds perfectly natural and true, whereas (26) sounds odd and false, despite the fact that practically every white bear that actually exists is a polar bear and thus eats seals, and the fact that the corresponding bare plural generics both sound fine and are true (25a, 26a).

(25) Polar bears, by virtue of being polar bears, eat seals.
(26) #White bears, by virtue of being white bears, eat seals.
(25a) Polar bears eat seals.
(26a) White bears eat seals.

This contrast follows naturally if the concept POLAR BEAR has a form such as (17), whereas WHITE BEAR has a form such as (24). This seems to be the case. As illustrated by (27), the concept WHITE BEAR does not provide the means for thinking about a single kind of bear. As such, white bears cannot be understood to have any properties by virtue of being white bears because to be a white bear is not to be a single kind of bear. WHITE BEAR simply provides the means for creating an equivalence class. It allows us to think about any kind of bear as long as it is white.

(27) Many kinds of bears can be white.
(28) #Many kinds of bears can be polar bears.

On the other hand, to be a polar bear is to be a single kind of bear (28), and thus a polar bear can have properties by virtue of being a polar bear, because to be a polar bear is to be some one kind of thing.

Representation of "Well-Formed Kinds" and the Interpretation of Indefinite Singulars

The differences in the formal structure of mechanisms (17) and (24) may also ground the distinction between so-called "well-established kinds" and kinds that are not "well-established kinds." This distinction is relevant to whether or not definite singular forms can be used to refer to a kind, for example, (29) and (30) (Carlson, 1980):

(29) The Coke bottle has a narrow neck.
(30) ??The green bottle has a narrow neck.

In the terminology being employed in the present chapter and Prasada and Dillingham (2006), the contrast is between *kinds* and *mere classes/types/equivalence classes*, with language and thought about kinds requiring the use of a mechanism such as (17), whereas talk and thought about mere classes requires use of a mechanism such as (24).

If this is correct, it suggests a refinement of our understanding of the distribution of indefinite singular generics. According to the proposal developed above, principled connections require a mechanism such as (17), and indefinite singulars can be interpreted as characterizing an arbitrary instance of a kind only when there is a principled connection between the kind and the characterizing property. Recall that Prasada and Dillingham (2009) found that (12), but not (13), may be interpreted as characterizing an arbitrary instance. This predicts that mere types, which involve a mechanism such as (24), should not support the generic use of indefinite singulars, because this sort of mechanism does not provide the means for representing principled connections. On the assumption that GREEN BOTTLE provides the means for thinking about a mere class, it should not be possible to use an indefinite singular to characterize green bottles. However, this prediction seems to be at odds with Krifka et al.'s (1995) judgment that (31) is acceptable.[11]

(31) A green bottle (usually) has a narrow neck.
(32) ??A green bottle has a narrow neck.
(33) Green bottles (usually) have narrow necks.
(34) Green bottles have narrow necks.

I suggest, however, that the present proposal is not in conflict with this point. To begin with, the acceptability of (31) seems to rely heavily on the presence of the 'usually' (32). The bare plural version, on the other hand, seems to be minimally affected by the presence/absence of 'usually' (33–34). These observations suggest that the indefinite singular may be used to characterize a mere class (e.g., GREEN BOTTLE) if it is interpreted as providing a "statistical characterization" (a characterization of what is usually the

---

11. Example 25a on page 11 of Krifka et al. (1995).

case). The contrast between (31) and (32) suggests that statistical characterization interpretations of indefinite singulars are possible only if an adverbial quantificational element such as 'usually' is explicitly present. Without this element, as predicted by the present account, indefinite singulars cannot characterize an arbitrary instance of a mere class.[12]

Arbitrary Instances and Arbitrarily Chosen Instances

This analysis raises a new problem. How could an arbitrary instance be given a statistical characterization? According to the proposal defended above, an arbitrary instance of a kind is a mechanism of thought that provides a nonquantificational source of generality. On the other hand, a statistical characterization is, of course, inherently quantificational. I'd like to argue that this apparent conflict is the result of not clearly distinguishing between an *arbitrary instance* of a kind and an *arbitrarily chosen instance* of a kind. An arbitrarily chosen instance of a kind is an actual instance of a kind, as in (35), unlike an arbitrary instance of a kind, as in (36). Furthermore, there can be indefinitely many arbitrarily chosen instances of a kind, as in (37), but there is only one arbitrary instance of a kind, as in (38).[13]

(35) Is this (pointing to a bottle) the arbitrarily chosen green bottle you were talking about?
(36) #Is this (pointing to a bottle) the arbitrary green bottle you were talking about?
(37) How many arbitrarily chosen green bottles do you want?
(38) #How many arbitrary green bottles do you want?

Each arbitrarily chosen instance of a kind, being an actual instance of a kind, either possesses or lacks any nonnecessary property of the kind. For example, each arbitrarily chosen dog either has four legs or does not have four legs. As such, it is not possible to characterize a nonspecific, arbitrarily chosen instance as either possessing [(39)] or lacking [(40)] a nonnecessary property of the kind. Instead, we must say what is usually the case, as in (41). This knowledge is necessarily quantificational. In contrast, we can specify how many legs an arbitrary dog has, as in (42), and it sounds odd to try to characterize what is usually the case, as in (44).

(39) #An arbitrarily chosen dog has four legs.

---

12. Indefinite singular generics will be possible for mere classes in the special cases where the characterizing property can be derived from characterizing properties of its constituents. For example, 'A green glass bottle casts a green bottle-shaped shadow' is possible because it derives from 'An X casts an X-shaped shadow' and 'X-colored glass casts X-colored shadows.'

13. This distinction can also serve to answer another criticism of arbitrary objects raised by Frege (Pelletier and Asher, 1997), namely, that mathematical proofs routinely make use of the notion of multiple arbitrarily chosen numbers or figures, even though there can be only one arbitrary instance of a kind.

(40) #An arbitrarily chosen dog does not have four legs.
(41) An arbitrarily chosen dog usually has four legs.
(42) An arbitrary dog has four legs.
(43) #An arbitrary dog does not have four legs.
(44) #An arbitrary dog usually has four legs.

This pattern of results is expected if the statements about an arbitrary instance of a kind reflect the knowledge embodied in the mechanism of thought that allows us to think about things as instances of kinds, as in (17), whereas statements about arbitrarily chosen instances of kinds reflect our knowledge of the statistical prevalence of the characterizing property.

Returning to the question concerning the distribution of generic indefinite singulars, it seems that simple indefinite singulars [(12)] characterize what an arbitrary instance is like and thus are limited to expressing principled connections, as in (17). Such statements can only be made about "well-established" kinds because they are grounded in a mechanism such as (17). Indefinite singular generics will also be possible for mere classes in the special case where the characterizing property can be derived from characterizing properties of its constituents (see footnote 12).

Explicitly statistical indefinite singulars, on the other hand, characterize what an arbitrarily chosen instance of a kind is likely to be like. Such statements may be made about arbitrarily chosen instances of mere classes, as in (31), or about properties of "well-established" kinds that lack a principled connection to a kind, as in (45):

(45) A barn is usually red.

Such generic statements are grounded in conceptual mechanisms other than (17). I turn now to very briefly consider what these mechanisms might be like.

Statistical Generics

The conceptual basis for statistical generics such as (5) is provided by a large enough proportion (majority?) of the representations of the specific instances of kinds generated by (17) and (24) having a specific value (e.g., red) for an aspect that is not determined by the kind of thing something is (e.g., color for barns). In this case, quantification is clearly relevant. Furthermore, some statistical principle must license the generalization. After all, (5) is a generic statement about barns rather than a statement about the prevalence of redness in a particular sample of barns. The statistical principle will determine the answer to questions such as how large the sample of barns must be, what proportion of the barns must be red, and what the distribution of the colors of nonred barns may be. There has been virtually no experimental work investigating the nature of the principle that licenses statistical generics such as (5); however, it is clear that it will make reference to the information concerning the number of factual connections encoded between the specific instances of a kind generated by (17) and (24) and the property in question.

Other Forms of Genericity

Yet other conceptual mechanisms are likely implicated in our understanding of generics such as (46), which can be true despite the fact that the vast majority of mosquitoes do not carry the West Nile virus.

(46) Mosquitoes carry the West Nile virus.

Leslie (2007; 2008) argues that (46) is an example of a type of generic that is licensed only if it involves an appalling, horrific, or striking property and if those instances that do not have the property are at least disposed to have the property. In line with the second condition, Prasada and Dillingham (2009) suggest that such cases may rely on the causal-essentialist mode of explanation and require representing a causal connection between being a mosquito or the essence of a mosquito[14] and the property of carrying West Nile virus. Because mosquitoes are rarely exposed to the difference-making causal conditions that allow them to develop the West Nile virus, very few mosquitoes actually carry the virus. Simply representing the causal connections between being a given kind of thing and having a given property seems to be sufficient for licensing generic statements such as (46). It seems that though either the representation of a causal or principled connection between a kind and a property can license a generic, only principled connections license expectations concerning the general prevalence of the property, as well as normative expectations concerning the presence of the property (there is nothing wrong with a mosquito that does not carry the West Nile virus) or formal explanations ("Mosquitoes, by virtue of being mosquitoes, carry West Nile virus" sounds odd).

Conceptual Representations and Genericity

Commonsense concepts provide us the means for thinking and talking about things from rich, intricate, and sometimes conflicting perspectives (Chomsky, 1996, 2000; McGilvray, 1999, 2005; Moravcsik, 1981, 1998, 2002; Pustejovsky, 1995). The mechanisms discussed in this chapter allow us to think about particular things as instances of kinds, as in (17), or as an instance of a class, as in (24). The former mechanism allows us to think of a kind either as a single thing or as the indefinitely many things that constitute the kind, or both at once. It does this by characterizing what an

---

14. The disjunction is to accommodate both the standard (e.g., Ahn et al., 2001; Bloom, 2000; Gelman, 2003) and minimalist versions of causal essentialism (Strevens, 2000, 2001)

arbitrary instance of the kind is like. As such, (17) provides an intrinsically general point of view from which we can think about things; (24) also provides an intrinsically general perspective from which to think about things, but, as discussed above, this perspective differs from the one provided by (17) in systematic ways. Viewing concepts as mechanisms or organs of the mind with specific structures that afford specific ways of thinking is an attempt to extend an approach to the study of mind that has proven very successful in studying other aspects of mind (Chomsky, 1975; Gallistel, 2007).

Some promising avenues of investigation suggested by the research reported here include (i) the possibility of capturing the fact that definite singulars seem to have more namelike behavior than do bare plurals (see chapter 2) in terms of differences between the ways in which the two types of linguistic expressions map onto mechanisms [(17) and (24)]; (ii) the possibility that the facilitatory role that early nouns seem to have in kind-relevant individuation and category formation may be due to there being a link between nouns and conceptual mechanisms such as (17) (Waxman, 2004; Xu et al., 2005); (iii) the possibility that many of the constraints on the definite generic may be understood in terms of whether the nominal picks out a single kind or picks out things of many kinds that satisfy a description [e.g., (27) and (28)]; (iv) the possibility of clarifying the suggestion that generics are developmentally primitive (Hollander et al., 2002; Leslie, 2006) in terms of the mechanism in (17); and (v) the possibility that children's very early ability to refer to both instances of kinds (Gelman et al., 2005) is a consequence of the fact that both these abilities may be rooted in the structure of the mechanism in (17).

More generally, an important question for future research is to investigate the representational structure implicated by other modes of commonsense explanation (Prasada and Dillingham, 2009) and how this structure is related to the structure implicated by the formal mode of explanation investigated in the research described here. Finally, a question of vital importance concerns how content domain-specific constraints on possible principled connections are incorporated into the content domain-general formal structure of kind concepts found in (17).[15] Such an account is needed in order to provide anything more than a vague sketch of how generic knowledge of the sort in (1) may be acquired.

Though the research described in this chapter barely begins to scratch the surface, it suggests that detailed investigation of our understanding of generics is likely to advance our understanding of the conceptual mechanisms

---

15. These mechanisms are domain-general in the sense that they allow us to think about kinds from all content domains. On the other hand, the mechanisms are domain-specific in that their structure is likely specific to representations within our conceptual systems. Thus, it is unlikely that mechanisms such as (17) will be found within our motor systems, though, of course, this is an empirical question.

available to commonsense conception. Conversely, it seems that it may be possible to understand certain otherwise unexplained aspects of the linguistic expressions of genericity in terms of the conceptual mechanisms underlying them. I expect the "Pelletier squish" which suggests a range of finely distinguishable generics (see Link, 1995) will provide valuable fodder for improving our understanding of both conceptual mechanisms and the nature of genericity for many years to come.

ACKNOWLEDGMENTS

I thank two anonymous reviewers for helpful comments. The research described here was supported by a grant PSC-CUNY 34 from the Profession Staff Congress of the City University of New York and by startup funds from Hunter College. It also received infrastructure support from Research Centers in Minority Institutions grant RR03037 from the National Center for Research Resources, National Institutes of Health, to the Gene Center at Hunter College.

REFERENCES

Ahn, W., Kalish, C., Gelman, S.A., Medin, D.L., Luhmann, C., Atran, S., Coley, J.D., and Shafto, P. (2001). Why essences are essential in the psychology of concepts. *Cognition*, 82, 59–69.

Bloom, P. (2000). *How Children Learn the Meanings of Words*. MIT Press, Cambridge, Mass.

Burton-Roberts, N. (1977). Generic sentences and analyticity. *Studies in Language*, 1, 155–196.

Carey, S. (1985). *Conceptual Change in Childhood*. MIT Press, Cambridge, Mass.

Carlson, G.N. (1980). *Reference to Kinds in English*. Garland Press, New York.

Carlson, G.N., and Pelletier, F.J. (eds). (1995). *The Generic Book*. Chicago: University of Chicago Press.

Chierchia, G. (1998). Reference to kinds across languages. *Natural Language Semantics*, 6, 339–405.

Chomsky, N. (1975). *Reflections on Language*. Random House, New York.

Chomsky, N. (1996). *Powers and Prospects: Reflections on Human Nature and the Social Order*. South End Press, Boston.

Chomsky, N. (2000). *New Horizons in the Study of Language and Mind*. Cambridge University Press, Cambridge, U.K.

Cohen, A. (2001). On the generic use of indefinite singulars. *Journal of Semantics*, 18, 183–209.

Dahl, O. (1975). On generics. In *Formal Semantics of Natural Language*, E. Kennen (ed.), pp. 99–111. Cambridge University Press, Cambridge, U.K.

Feigenson, L., and Carey, S. (2005). On the limits of infants' quantification of small object arrays. *Cognition*, 97, 295–313.

Fodor, J.A. (1998). *Concepts: Where Cognitive Science Went Wrong*. Oxford University Press, New York.

Gallistel, C.R. (2007). Learning organs. English original of L'apprentissage de matières distinctes exige des organes distincts. In *Cahier n° 88: Noam Chomsky*, J. Bricmont and J. Franck (eds.), pp. 181–187. L'Herne, Paris.

Gelman, R. (1990). First principles organize attention to and learning about relevant data: Number and the animate-inanimate distinction as examples. *Cognitive Science*, 14, 79–106.

Gelman, S. (2003). *The Essential Child: Origins of Essentialism in Everyday Thought*. Oxford University Press, London.

Gelman, S.A., and Wellman, H.M. (1991). Insides and essences: Early understanding of the non-obvious. *Cognition*, 38, 213–244.

Gelman, S.A., Chesnick, R.J., and Waxman, S.R. (2005). Mother-child conversations about pictures and objects: Referring to categories and individuals. *Child Development*, 76, 1129–1143.

Gergely, G., Nadasdy, Z., Csibra, G., and Biro, S. (1995). Taking the intentional stance at 12 months of age. *Cognition*, 56, 165–193.

German, T.P., and Johnson, S. (2002). Function and the origins of the design stance. *Journal of Cognition and Development*, 3, 279–300.

German, T.P., Niehaus, J.L., Roarty, M.P., Giesbrecht, B., and Miller, M.B. (2004). Neural correlates of detecting pretense: Automatic engagement of the intentional stance under covert conditions. *Journal of Cognitive Neuroscience*, 16, 1805–1817.

Gopnik, A., and Meltzoff, A.N. (1997). *Words, Thoughts, and Theories*. MIT Press, Cambridge, Mass.

Greenberg, Y. (2003). *Manifestations of genericity*. Routledge, New York.

Hollander, M.A., Gelman, S.A., and Star, J. (2002). Children's interpretation of generic noun phrases. *Developmental Psychology*, 6, 883–894.

Inagaki, J., and Hatano, G. (2002). *Young Children's Naive Thinking about the Biological World*. Psychology Press, New York.

Johnson, S.C., and Solomon, G.E.A. (1997). Why dogs have puppies and cats have kittens: The role of birth in young children's understanding of biological origins. *Child Development*, 68, 404–419.

Kahneman, D., Treisman, A., and Gibbs, J.B. (1992). The reviewing of object files: Object-specific integration of information. *Cognitive Psychology*, 24, 175–219.

Keil, F.C. (1989). *Concepts, Kinds, and Conceptual Development*. MIT Press, Cambridge, Mass.

Keil, F.C. (1994). The birth and nurturance of concepts by domains: The origins of concepts of living things. In *Mapping the Mind: Domain Specificity in Cognition and Culture*, L. Hirschfeld, and S. Gelman (eds.), pp. 234–254. Cambridge University Press, New York.

Kelemen, D. (1999). The scope of teleological thinking in preschool children. *Cognition*, 70, 241–272.

Krifka, M., Pelletier, F.J., Carlson, G.N., ter Meulen, A., Link, G., and Chierchia, G. (1995). Genericity: An introduction In *The Generic Book*, G.N. Carlson and F.J. Pelletier (eds.), pp. 1–124. University of Chicago Press, Chicago.

Kuhlmeier, V.A., Bloom, P., and Wynn, K. (2004). Do 5-month-old infants see humans as material objects? *Cognition*, 94, 95–103.

Lawler, J. (1973). *Studies in English generics*. University of Michigan Papers in Linguistics 1:1. University Press, Ann Arbor, Mich.

Leslie, A.M. (1994). ToMM, ToBY, and Agency: Core architecture and domain specificity. In *Mapping the Mind: Domain Specificity in Cognition and Culture*, L.A. Hirschfeld and S.A. Gelman (eds.), pp. 199–148. Cambridge University Press, New York.

Leslie, S.J. (2007). Generics and the structure of the mind. *Philosophical Perspectives, 21*, 375–403.

Leslie, S.J. (2008). Generics: Cognition and acquisition. *Philosophical Review, 117*, 1–47.

Link, G. (1995). Generic information and dependent generics. In *The Generic Book*, G.N. Carlson and F.J. Pelletier (eds.), pp. 358–382. University of Chicago Press, Chicago.

Lombrozo, T., and Carey, S. (2006). Functional explanation and the function of explanation. *Cognition, 99*, 167–204.

Macnamara, J., and Reyes, G.E. (1994). Foundational issues in the learning of proper names, count nouns and mass nouns. In *The Logical Foundations of Cognition*, J. Macnamara and G.E. Reyes (eds.), pp. 144–176. Vancouver Studies in Philosophy, Volume 4. Oxford University Press, Oxford.

McGilvray, J. (1999). *Chomsky: Language, Mind, and Politics*. Polity Press, Cambridge, U.K.

McGilvray, J. (2005). Meaning and creativity. In *The Cambridge Companion to Chomsky*, J. McGilvray (ed.), pp. 204–222. Cambridge University Press, Cambridge, U.K.

Medin, D.L., and Shaffer, M.M. (1978). Context theory of classification learning. *Psychological Review, 85*, 207–238.

Moravcsik, J.M.E. (1981). How do words get their meanings. *Journal of Philosophy, 78*, 5–24.

Moravcsik, J.M.E. (1998). *Meaning, Creativity, and the Partial Inscrutability of the Human Mind*. CSLI Publications, Stanford, Calif.

Moravcsik, J.M.E. (2002). Chomsky's New Horizons: Review of Noam Chomsky, *New Horizons in the Study of Language and Mind*. *Mind and Language, 17*, 303–311.

Opfer, J.E., and Gelman, S.A. (2001). Children's and adult's models for predicting teleological action: The development of a biology-based model. *Child Development, 72*, 1367–1381.

Pelletier, F.J., and Asher, N. (1997). Generics and defaults. In *Handbook of Logic and Language*, J. van Benthem and A. ter Meulen (eds.), pp. 1125–1177. North Holland, Amsterdam.

Prasada, S. (2000). Acquiring generic knowledge. *Trends in Cognitive Sciences, 4*, 66–72.

Prasada, S., and Dillingham, E.M. (2006). Principled and statistical connections in common sense conception. *Cognition, 99*, 73–112.

Prasada, S., and Dillingham, E.M. (2009). Representation of principled connections: A window onto the formal aspect of common sense conception. *Cognitive Science, 33*, 401–448.

Prasada, S., Salajegheh, A., Bowles, A., and Poeppel, D. (2008). Characterizing kinds and instances of kinds: ERP reflections. *Language and Cognitive Processes, 23*, 226–240.

Pustejovsky, J. (1995). *The Generative Lexicon*. MIT Press, Cambridge, Mass.

Pylyshyn, Z. (1989). The role of location indexes in spatial perception: A sketch of the FINST spatial-index model. *Cognition, 32*, 65–97.

Quine, V.W.O. (1987). *Quiddities: An Intermittently Philosophical Dictionary*. Belknap Press, Cambridge, Mass.

Scholl, B. (2001). Objects and attention: The state of the art. *Cognition, 80*, 1–46.

Smith, L.B., and Samuelson, L.K. (1997). Perceiving and remembering: Category stability, variability, and development. In *Knowledge, Concepts, and Categories*, K. Lamberts and D. Shanks (eds.), pp. 161–195. MIT Press, Cambridge, Mass.

Springer, K., and Keil, F.C. (1991). Early differentiation of causal mechanisms appropriate to biological and nonbiological kinds. *Child Development, 62*, 767–781.

Strevens, M. (2000). The essentialist aspect of naïve theories. *Cognition, 74*, 149–175.

Strevens, M. (2001). Only causation matters: Reply to Ahn et al. *Cognition, 82*, 71–76.

Tennant, Neil (1983). A defense of arbitrary objects. *Proceedings of the Aristotelian Society, suppl. vol. 57*, 79–89.

Waxman, S.R. (2004). Everything had a name, and each name gave birth to a new thought: Links between early word learning and conceptual organization. In *Weaving a Lexicon*, D.G. Hall and S.R. Waxman (eds.), pp. 295–335. MIT Press, Cambridge, Mass.

Whittlesea, B.W.A. (1997). The representation of general and particular knowledge. In *Knowledge, Concepts, and Categories*, K. Lamberts and D. Shanks (eds.), pp. 335–370. MIT Press, Cambridge, Mass.

Xu, F., Cote, M., and Baker, A. (2005). Labeling guides object individuation in 12-month-old infants. *Psychological Science, 16*, 372–377.

4

Are All Generics Created Equal?

FRANCIS JEFFRY PELLETIER

Generic statements of the sort under consideration in this article are those such as

(1) a. Potatoes contain vitamin C.
　　b. Basketball players are tall.
　　c. Predatory animals have sharp teeth.
　　d. Birds fly.

Such genericity is a feature of entire sentences (or at least of entire clauses) and is not attributable to any subpart of the sentence, such as to the subject noun phrase or to the verb phrase. It is instead somehow an interaction between the two. This sets it apart from another type of genericity, which is attributed to the noun phrase referring to an abstract genus (or kind) and not as making a statement about the individual members of that genus, as in

(2) a. The potato was first grown in South America.
　　b. Dodos are extinct.
　　c. Mosquitos are common in northern Canada.
　　d. Man landed on the moon in 1969.

Although there are various commonalities between the two types, and certain relations that hold between them, we are here interested only in the first phenomenon, which was designated "characterizing genericity" in Krifka et al. (1995).

One of the logically, semantically, and psychologically interesting features of generic statements is that they tolerate exceptions. That is, they can be true despite the fact that there are some particular instances that do not possess the generically predicated property. For example, (1a) is true despite some spoiled potatoes, and (1d) is true despite ostriches, emus, penguins, kiwis, cassowaries, and the like. This feature is covered extensively in Krifka et al. (1995) and is simply assumed in the present article.

Characterizing genericity can be expressed not only by the "bare plural" formulation given in (1) but also with definite [(3a)] and indefinite [(3b)] count noun phrases and by mass nouns ["bare singulars," (3c)]. In addition, the subject term can be a proper name, [(3d)], although in that case, the characterizing genericity phenomenon is usually called "habitual."

(3) a. The potato is highly digestible.
    b. A snake is a reptile.
    c. Snow is white.
    d. Fred has wine with dinner.

Not only can the item that is being claimed "to generically have a property" occur in the subject position, as all these examples suggest, but it can also occur in object positions, and indeed, there can be more than one in a sentence, as these examples show:

(4) a. Cats chase mice.
    b. Canadians are taller than Mexicans.
    c. Paul smokes cigars after meals.
    d. Italians love pasta.

Many of these types of generics raise special puzzles within formal semantics, and their position within the pantheon of generics is not yet settled. For example, it was pointed out in Lawler (1973) that the indefinite formulation as in (3b) works only when the predicate is somehow "essential" to the subject term. And comparatives such as (4b) seem to call for some special treatment that involves comparisons between averages (Carlson and Pelletier, 2002). Habituals such as (3d) and (4c) seem to generalize over events, rather than over individuals as the others do. Furthermore, mass terms themselves bring a set of unique problems (Pelletier and Schubert, 2003), and it might be best to await a semantic treatment for them before tackling generics such as (3c). Finally, there might be important semantic differences between definite generics as in (3a) and the bare plural formulations in (1), so we perhaps should reserve them for separate treatment.

Thus, the present article discusses only variants of the bare plural formulation, and then only in subject position of simple sentences that do not involve a relational statement. We furthermore reserve for separate treatment the types that are mentioned in (2), even when they employ bare plurals. (The common attitude is that the types of predicates in these sentences are of a different sort than the more traditional generic predicates in (1), but I do not wish to engage that discussion here.)

But even within just the indicated type of generics, one can find distinct formulations. For example, we might see (1c) stated in any of these ways (and others):

(5) a. Predatory animals have sharp teeth. (BP)
    b. Most predatory animals have sharp teeth. (Q)
    c. The typical predatory animal has sharp teeth. (M)
    d. Predatory animals usually have sharp teeth. (Adv-1)
    e. Predatory animals normally have sharp teeth. (Adv-2)

The (a) version is called the "bare plural" formulation (BP), the (b) version is the "quantificational" formulation, the (c) version is the "noun modifier" version, and the final two are labeled as two different "adverbial" formulations. Other forms can be found in the literature, but these are representative of the variety that are mentioned. Mostly, authors will mention all of these as merely alternative formulations of the same semantic item—that is, they are claimed to have the same generic force or meaning. There has been, however, an opinion that there is a difference in the ways that these generic sentences have been given a theoretical footing. This view is commonly attributed to Greg Carlson (see, e.g., Carlson, 1980, 1982, 1995), who calls them "the inductivist approach" and "the rules and regularities approach." The former approach has a driving intuition that our generic sentences basically express inductive generalizations, where the basis of the generalization is some observed set of instances. The idea is that, after observing a number of instances of predatory animals with sharp teeth, a sentence such as (5a) is generated as a covering description. In describing the philosophical backdrop of this view of generic sentences, Carlson (1995, p. 225) says "the most natural bedfellows of this approach would be empiricists, verificationists, and nominalists of varying stripes." A different background view is the latter approach, which does not hold that generics are truly asserted on the basis of *any* array of instances, but rather that they depend for their truth or falsity upon whether or not there is a causal organization within the world that corresponds to them. The philosophical perspective of those who would adopt this view are those who, in Carlson's words (1995, p. 225), "admire properties and propositions as real entities, . . . as would many realists."

This chapter is an investigation into the status of these sorts of claims. There are two related aspects: first is the view, taken by almost all writers on generics, whether within philosophy or linguistics or artificial intelligence, that the sentences in (5) all express genericity equally; and second is the view expressed by Carlson that there is a difference between two theoretical attitudes toward the grounds for the truth of generic sentences. Note that Carlson's view is that each of the two theoretical viewpoints would hold that all the sentences in (5) express genericity equally; it is just that they differ in what genericity amounts to.

This study aims to challenge the view that all genericity is viewed the same way by ordinary speakers of natural language (in this case, English). Intuitively, a sentence such as (5b) ought to be a paradigm case of the inductivist approach since 'most' seems clearly to call for some sort of explicit comparison of the number of predatory animals with versus without sharp teeth. And again, the version worded as (5a) seems most straightforwardly to embody the rules and regularities approach, since we seem to be appealing to the very kind itself, PREDATORY ANIMALS, and its regulatory properties. The other versions in (5) can be seen as going either way, or perhaps even changing their interpretations depending on the context or example under discussion. I show that there are at least two different types of generic statements and that they are lexically differentiated. A natural interpretation might be that the two types correspond to Carlson's two viewpoints, and that each viewpoint finds itself most clearly represented by different ones of the sentences in (5). But that is a further interpretation to put upon the present findings. What we can say is that our findings are consistent with the view that the two different background interpretations for characterizing generics find expression in two syntactically different sentence types.

Background to Default Reasoning

Default reasoning occurs whenever the evidence available to the reasoner does not guarantee the truth of the conclusion being drawn, that is, does not deductively force the reasoner to draw the conclusion under consideration ("force" in the sense of being required to do it if the reasoner is to be logically correct). Nonetheless, the reasoner does draw the conclusion, and is correct in doing so. For example, from the statements 'Linguists typically speak at least three languages' and 'Kim is a linguist,' one might draw the conclusion, by default, that 'Kim speaks at least three languages.' What is meant by the phrase "by default" is that we are justified in making this inference because we have no information that would make us doubt that Kim was covered by the generalization concerning linguists or would make us think that Kim was an abnormal linguist in this regard. Of course, the inference is not deductively valid: it is possible that the premise could be true and the conclusion false. So, one is not forced to draw this conclusion in order to be logically correct. Rather, it is the type of conclusion that we draw "by default"—the type of conclusion we draw in the ordinary world and ordinary circumstances in which we find ourselves.

Default reasoning is nonmonotonic; that is, adding new premises can make us withdraw previously generated conclusions without withdrawing any of the previous premises. For example, were we to add to our list of premises the further fact that Kim graduated from NewWave University, which we know has revoked all language requirements, we then would wish to withdraw the earlier conclusion, even though we would not withdraw any of our other premises.

Default reasoning makes its appearance in many academic disciplines. Pelletier and Elio (2005) cite examples from ethics, philosophy of science, conditional and counterfactual logics, relevance logics, studies of prototypical and stereotypical schemata, causal reasoning, medical and fault diagnosis, reasoning in the social sciences, judgments under uncertainty, reasoning involving (Gricean) implicatures, lexical defaults in linguistics, "natural" logic (as pursued, e.g., by Lakoff, 1972, 1973), knowledge representation in artificial intelligence (AI), and some argumentation about the nature of cognition in cognitive science. In fact, the leading examples of all these fields are what would be called "characterizing generics" by Krifka et al. (1995).

Lifschitz (1989) published a set of "benchmark problems" that all formalisms for default reasoning are supposed to follow. These examples covered a wide variety of different applications; the ones I am interested in for the present study form the first part of his paper: three of the four problems he called "Basic Default Inference." Here are the three, as presented in Lifschitz (1989):

Benchmark 1
Blocks A and B are heavy.
Heavy blocks are normally located on this table.
A is not on this table.
Therefore, B is on this table.

Benchmark 2
Blocks A and B are heavy.
Heavy blocks are normally located on this table.
A is not on this table.
B is red.
Therefore, B is on this table.

Benchmark 3
Blocks A and B are heavy.
Heavy blocks are normally located on this table.
Heavy blocks are normally red.
A is not on this table.
B is not red.
Therefore, B is on this table.

Each of these problems concerns two objects governed by one or more default rules. Additional information is given to indicate that one of the objects (at least) does not follow one of the default rules. We refer to this as the exception object (for that default rule). The problem then asks for a conclusion about the remaining object, which we refer to as the object-in-question. For all these problems, Lifschitz endorses the conclusion that the object-in-question (block B) obeys the default rule concerning location. According to the collective wisdom of researchers into nonmonotonic theories, the existence of an exception object for a default rule (benchmark 1), or additional information about that exception object, should have no bearing on a conclusion drawn about any other object when using that rule. Extra information about the object in question itself (e.g., block B's color in benchmark 2) should also have no bearing on whether a default rule about location applies. And being an exception object for some other default rule should have no bearing on whether it does or does not follow the present default rule (benchmark 3). I call these the "AI-approved answers" to the benchmark problems.

The default rules used in Lifschitz's problems are in fact the Adv-2 versions of the variant ways of stating generics given in (5), so we might ask whether people treat this version of a generic statement the same or different from the other versions of the same content. One problem with using Lifschitz's problems directly is that the generic statement (i.e., the default rule) is what might be called a nonce-generic, that is, a spur-of-the-moment invention. We should wish to investigate generics that are more "stable" in their interpretations. But in doing so we need to avoid well-known generic truths (and falsehoods), since this knowledge could affect subjects' answers about whether or not the conclusion follows. For instance, we should not ask about whether 'Tweety the penguin flies' follows from 'Birds normally fly' and 'Penguins are (always) birds,' since it is a piece of stored knowledge about penguins that they do not fly. Hence, subjects would not be calling upon any reasoning module to answer such a question, but would instead just be engaged in some sort of memory look-up.

Some theorists have also thought that generic statements about natural kinds might be internalized differently by people than are ones about artifacts, since the former maybe are governed by "natural laws" while the latter could just have "accidental facts" be true about them. (Or maybe the other way around: things in the artifactual world are made for a purpose, whereas things in the natural world exhibit just whatever behavior happens to hold for them.)

For these reasons, I made up example generics using both made-up "natural" kinds and made-up "artifactual kinds." A cover story was used to introduce the kinds and to make clear whether or not they were natural or artifactual kinds. The story went on to characterize the kinds along the lines of the benchmark problems, by giving the "generic information," and the subjects were asked to evaluate how well the story supported the benchmark conclusion. The goal then was to investigate whether the five different ways of giving the generic information mentioned in (5) above would affect how strongly the subjects thought the conclusion followed.

Method and Results

Two experiments were conducted: the first was a largish pilot study that attempted to determine whether there were likely to be detectable effects in asking questions about the benchmark problems when using only slightly syntactically different formulations. Despite the lack of controls for various factors, the results of this pilot study are interesting for their descriptive data. A follow-up experiment attempted to measure more carefully the factors involved. Both experiments used the same general type of setup, to which we now turn.

A Pilot Study

Materials in Pilot Study

As remarked above, subjects were presented with "cover stories" that characterized some kind—natural or artifactual—that they would not have heard about before (since they were made up for the experiment). These cover stories were described in the instructions to subjects as "paragraphs taken from newspaper or magazine stories," and they described various features of these kinds. The things that were mentioned about the kinds mapped directly to the features mentioned in the original benchmark problems, and the subjects were then asked to describe the extent to which they thought the conclusion (which also mapped directly to the benchmark problem) followed from the information presented.

The pilot experiment was of a $3 \times 2 \times 6$ design: three benchmarks (BM1, BM2, BM3), two types of kinds involved (natural vs. artifactual: NK, AK), and six different types of information. Five of these types were generic information (BP, Q, M, Adv-1, Adv-2), and one further version, Inv, used an existential quantifier rather than a type of generic (e.g., 'Some predatory animals have sharp teeth' rather than 'Predatory animals usually have sharp teeth'). The intent was that this sort of information—rather than a generically stated premise—should make the subjects decide that the conclusion *did not* follow, and that the argument was invalid. Generally speaking, this was to serve as a check that the subjects really were attending to the content of the generic information premise of the problems.

A sufficient number of different contents were constructed so that no subject received the same content with only a different version of the generic premise. For example, if a subject answered the BM1.NK.BP problem with content that dealt with (made-up) desert plants, then that subject would be given a BM1.NK.Q problem with, say, content dealing with (made-up) ocean current types, and a BM1.NK.M problem having planets as content, and so forth. This was enforced so as to prevent boredom and familiarity from becoming factors in subjects' answers to problems that differed only in the linguistic form of their generic information. The different contents of problems were balanced so that other subjects received their BM1.NK.BP problem about ocean currents, their BM1.NK.Q problem about planets, and so forth.

Subjects were presented with these "newspaper stories," each on a separate sheet of paper, and were to answer according to the strength that they thought the story supported the conclusion by marking a location on a 1–7 scale. The BM1.NK.M problem using the "conifer" story, the BM2.AK.Q problem using the "medieval musical instruments" story, and the BM1.AK.Inv (invalid) problem using "pillows" are given in the appendix.

Some of the problems given to subjects had the "AI-approved" answer on the left side (low numbers) of the answer scale, while others had it on the right side (high numbers) of the answer scale. In the results reported here, all scores have been normalized so that the "AI-approved" answers are high. Intuitively, then, a score greater than 4 (after normalization) is the AI-approved answer, scores less than 4 say that the object-in-question follows the exception object, and scores of about 4 say that one can't choose between them.

Had this experimental design been carried out correctly, all subjects would therefore have been given 36 problems to solve, with their order randomized among subjects. But due to an error in the collocation of the answer sheets, not all subjects received every condition. Subjects took between 30 and 55 minutes to answer the set of problems, averaging about 45 minutes.

Subjects in Pilot Study

There were 72 subjects in the pilot study, 41 female and 31 male, all native speakers of English. Their average age was slightly greater than 22 years. They were paid $10 for participating and were drawn from the population of Simon Fraser University. Most were students (both graduate and undergraduate), but 10 of them were visitors to the campus during a summer break.

Results of Pilot Study

Although the design of the experiment, as well as the error in collocation of some answer sheets and a typo in one group of problems, did not allow for detailed analysis of the data, there are nonetheless some very interesting descriptive statistics that can be stated. First, table 4.1 shows scores from the invalid versions of the six benchmark × kind problems contrasted with the averages of the other five versions of the same problem (the versions with some form of generic premise).[1] All these differences are significant (ignoring the issue with BM3.AK). Thus, subjects did, in fact, attend to the generic aspect of the premises in these sorts of problems, and did recognize that they were different from the similar premise with an existential quantifier in it. I take this to show the general viability of this direction of research.

Table 4.2 gives overall ranking of the three different benchmark problems, averaged across all the different kind and linguistic types (except the

---

1. An unfortunate but unnoticed typo in the BM3.AK.Inv problem made it not be appropriate as a contrast problem.

TABLE 4.1 Benchmark problems: Invalid arguments versus all valid generics averaged

Type	BM1.AK	BM1.NK	BM2.AK	BM2.NK	BM3.AK	BM3.NK
Invalid	2.94	3.80	4.92	3.82	*	4.21
Generics	5.75	5.79	5.72	5.36	5.17	5.23

invalids, of course). Note that each is significantly higher than the 4.0 "can't tell" midpoint score, and so each are even further from the "follows the exception object" answer. Therefore, people do in fact follow the AI-approved answer to benchmark questions, no matter what sort of generic formulation is used in the premise.[2]

Table 4.2 also shows that the overall scores for BM1 and BM2 are not significantly different from one another, but they are both significantly different from BM3. Recall that BM3 has the form 'A is an F; F's are usually G; F's are usually H; A is not an H; Is A a G?' In related previous research (Elio and Pelletier, 1993, 1996; summarized in Pelletier and Elio, 2002, 2005), we used a similar methodology with the purpose of investigating whether subjects treated the three benchmark problems the same. In those studies, unlike the present ones, we only used the Adv-1 ('usually') formulation, and we used nonce generics such as 'books required for my course,' rather than the made-up but "real" generics that we used in the present studies. Those studies also found that BM1 and BM2 were treated the same by subjects, but that BM3 was different. The AI-approved answer is 'yes, A is a G,' but we found that subjects were significantly less likely to say this than in BM1 and BM2. Various further manipulations (Elio and Pelletier, 1996) concerning the degree of similarity of the "extra property" mentioned in benchmark 3 to a presumed reason the exception object violated the default rule led us to a notion of "explanation-based default reasoning." In turn, this led us to speculate in our later summaries that an increased number of violated generic statements (adding 'F's are usually J; A is not a J,' etc.)

TABLE 4.2 Average "normalized" scores on benchmark problems: number of answers

Type	BM1	BM2	BM3	Total
Average	5.77	5.73	5.27	5.59
N	463	454	448	1,365

2. The different numbers of subjects in the different conditions illustrates the errors in collocation of the answer sheets and the problem with one set of answers. Nonetheless, it shows that there were large numbers of responses for each condition.

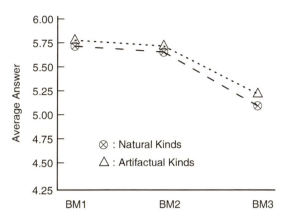

FIGURE 4.1 Benchmark × kind interaction

would make subjects less likely to give the AI-approved answer. We called this a second-order default ("The more defaults that are violated, the more likely to violate defaults"), or the "rotten egg principle" ("Once a rotten egg, always a rotten egg"), which seems to be endorsed by the results of the present pilot study.

Figure 4.1 gives the averages across all conditions (except Inv) for artifactual versus natural kinds. Although the AK condition seems to be slightly higher than the NK condition throughout all benchmark problems, the difference is not significant. Surprisingly, there was no main effect for linguistic type of generic. The averages for the different linguistic types across benchmark × kind are given in table 4.3.

There were between 260 and 300 answers for each of the types in table 4.3, and the differences were close to being significant. Using the Tukey HSD (Honestly Significant Difference), a difference of 0.25 would justify significance at the .05 level. Given the missing data in the cells, a between-subjects comparison had to be carried out here, and it seems plausible to think that a within-subjects set of comparisons could more cleanly give appropriate data. The group {Q,M,Adv-2} versus the group {BP,Adv-1} is very close to significantly different. (However, from the point of view of an intuitive difference, one might find it strange that bare plural generics and 'usually'-generics should be so close.)

Given these suggestive results, a more carefully controlled experiment was developed.

TABLE 4.3 Linguistic types, across all conditions

Type	Q	M	Adv-2	BP	Adv-1
Mean	5.41	5.42	5.45	5.64	5.65

ARE ALL GENERICS CREATED EQUAL?

Main Experiment

Materials in Main Experiment

The materials used in this experiment were a subset of those used in the pilot study, with some alterations discussed below. They were again the benchmark problems presented as "paragraphs taken from newspaper or magazine stories" used in the pilot study. The major design change was to eliminate two of the linguistic mode forms, the modifier (M) and the second adverb (Adv-2) forms, so as to leave only four syntactically different formats. One of these four was the "invalid" (Inv) form, so we tested only three different ways to say generics: the bare plural form (BP), the quantificational form (Q), and an adverbial form (Adv). (The adverbial form in this study was 'usually.') The experiment now has a $3 \times 2 \times 4$ design, so there are 24 different conditions being tested, and all subjects were tested in all conditions. As before, no subject had the same content in any two test conditions. Another change was that the Inv problems more closely imitated the generic information premise of the regular benchmarks, rather than being a specially constructed problem. Also, some of the exact wording for some problems was changed as a result of comments obtained from participants in the pilot study, and the lengths of some of the problems were altered so that they were all closer in length to each other.

Other changes involved using a randomized block design. As before, it is impossible to measure whether the different contents that a subject gets (when answering different levels of the linguistic mode for the same benchmark × kind problem) produces an effect. But the effects of different factors we measured here are based on within-subjects measurements.

Other features of the experiment, such as having some questions with a 1-answer meaning "the AI-approved answer" and others with a 7-answer having that meaning (and then normalizing so that 7 always means the AI-answer when we report the data), are the same in the two experiments. This being a shorter task, subjects almost always were done within 30 minutes.

Subjects in Main Experiment

There were 108 subjects in the main experiment, 65 female and 43 male, all native English speakers. One subject (a female) had to be eliminated owing to missing data on the answer sheets. Their average age was just over 21.5 years. Again, as in the pilot study, they were paid $10 for participating and were drawn from the population of Simon Fraser University. Unlike the pilot study, all these were undergraduate students. None of these subjects had participated in the pilot study.

Results of Main Experiment

As in the pilot experiment, subjects reliably distinguished the formulations with generic information—whether they were phrased 'Birds fly,' 'Most birds fly,' or 'Birds usually fly'—from the invalid, existential ones such as 'some birds fly.' And they can reliably do this even with novel types of entities, not just birds, and in complex scenarios. Figure 4.2 shows the means for the four linguistic modes (including the Inv mode). The Inv mode is significantly lower than the three versions of generics.[3]

However, the presence of an Inv premise affects the benchmark problems in a different way than do the generic premises, as figure 4.3 shows. When averaged over all the various generics, BM1 and BM2 are not significantly different from one another, but they are significantly different from BM3. But in the Inv case, the three problems are not significantly different from one another.

Once again, as in the pilot study and in the earlier work of Elio and Pelletier (1993, 1996), these data show that benchmark 3 is indeed treated quite differently from the other two benchmark problems when using generic premises, with subjects finding it a considerably worse argument when there is an "irrelevant" generic truth that the object-in-question disobeys, even though the conclusion does not mention that violation.

Having therefore shown that the various generic formulations actually tap something different than the invalid, existential formulation of the same

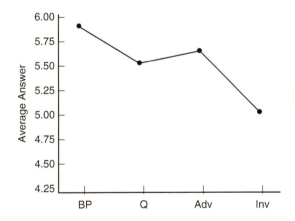

FIGURE 4.2 Averages of the four linguistic forms

---

3. All claims of of significant and nonsignificant differences are determined by a three-way repeated-measures analysis of variance, with significance set at $\alpha = .01$. When testing for factors with more than two levels, the Mauchly test of sphericity was run, and when it was not met, the Huynh-Feldt correction was used.

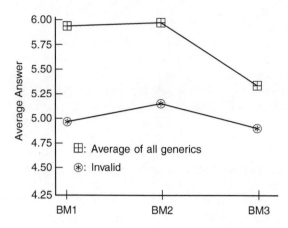

FIGURE 4.3 How the benchmark problems treat generics versus Inv

premise, we turn our attention to the way natural versus artifactual kinds affect subjects' reasoning (figure 4.4). Confirming the result of the pilot study, there is no significant effect for natural versus artifactual kinds in the three generic types of information. However, with the invalid versions there *is* an effect, with subjects recognizing the invalidity more easily when the kind in question is an artifact. The fact that natural–artifactual distinction affects nonmonotonically invalid arguments but not nonmonotonically valid arguments is itself a very interesting phenomenon for which I have no clear explanation. Certainly this deserves further study.

Turning our attention now to the way the three different benchmark problems deal with the linguistically different generic formulations, we see

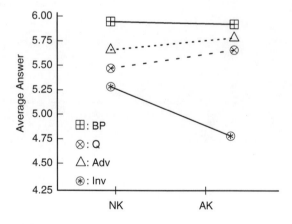

FIGURE 4.4 Linguistic form with different content-types

72 KINDS, THINGS, AND STUFF

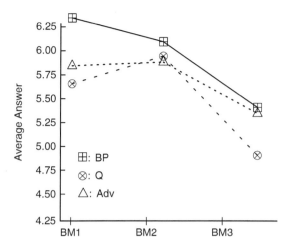

FIGURE 4.5 Linguistic form at different problems

in figure 4.5 that the problems treat the formulations differently. (i.e., there is an interaction between benchmark problem and linguistic form.) In benchmark 1, the BP formulation is significantly different from the Q and the Adv formulations, which are not significantly different from each other. In benchmark 2, none are significantly different from one another. And in benchmark 3, the Q formulation differs significantly from the others, which do not differ from each other. It therefore seems clear that the three linguistic formulations are tapping different aspects of generic information and that their differing types of information come to the fore in the different problems.

Finally, let us return to the overall data, which includes the Inv linguistic mode, and consider the magnitude of effect of the various factors that were found to be significant in the main study. These magnitudes are estimated by the proportion of variance in the subjects' judgment of validity that is explained by its relationship with one of these factors, after the effects of the blocking factor (subjects) and the interactions of the blocking factor with other factors have been removed. Thus, we estimated the size of the effect of problem type (i.e., which benchmark problem), after removing the blocking factor and its interactions. Similarly, we can estimate the magnitude of the effect of the linguistic mode on the judgment of validity, the magnitude of the effect of the interaction of content kind (natural vs. artifactual) with linguistic mode on the judgment of validity, and the the magnitude of the effect of the interaction of problem type with linguistic mode on the judgment of validity. Table 4.4 gives these results.

TABLE 4.4   Estimated Magnitude of Effect

Factor	Magnitude
Linguistic mode	55.6%
Problem type	30.4%
Problem × mode	06.5%
Problem × kind	04.9%
Other	02.6%

Discussion

I should reemphasize that the manipulation investigated in this study actually yields interpretable results. Informal discussions with other researchers have revealed that many of them believe that whatever semantic differences there might be among the various ways of expressing generic statements, they are extremely minor and easily overwhelmed by all sorts of other factors. And it is thought that tests involving the vague notion of default reasoning and asking for judgments of the validity of arguments will simply invoke too coarse a methodology for testing these minor differences. But the results of the experiments here are robust, both in the pilot study and in the main study: there is a clear difference between generically presented information and existentially presented information, and the subjects reliably detect this difference. Furthermore, there are detectable significant differences within the various linguistically different generic formulations.

The results, both from the pilot study and from the main study, also demonstrate that the use of a natural kind versus an artificial kind does not affect subjects' judgments concerning the validity of default argumentation that employs these kinds in generic statements. This might be somewhat surprising to those who have views concerning the way that natural-kind information comes into our consciousness as opposed to the way artificial-kind information comes in. But we also saw that this distinction *does* make a difference in the perception of *invalidity* of default arguments. Now, it is not at all clear why the natural–artifactual distinction should play a role with invalid (existential) premises but not with generic premises in these sorts of default reasoning. The answer could lie in the idea that people think of artifacts as created for a purpose and their properties as leading to that purpose and therefore having certain regularities, while natural items are simply "presented" to an observer and hence no purpose for their properties is clear. Or, perhaps natural items are assumed to have regularities that obey natural laws, while artifacts are created without any concern for natural regularities. These speculations are suggestive, each in its own way. But such explanations require more clear rationales and a closer tie to theories of learning (involving how kinds are learned), and to how default reasoning proceeds from these learned kinds, before the present results can be said to support one or the other of the speculations.

It is also heartening (for me) to note that both the pilot study and the main study supported our earlier finding (Elio and Pelletier, 1993, 1996) that benchmark problem 3 is significantly different from the others. A look at figure 4.3 shows that the average of the generics on benchmark 3 is approaching the level of the responses to the invalid versions of the benchmark problems. So, not only do subjects treat benchmark 3 differently from the other two, but in fact they seem to view it as (almost) invalid. The earlier studies did not conclude this, but rather that subjects saw the increase in the number of violated generic traits as being more and more telling against the AI-approved answer. But since there were no results for invalid default problems, we could not make the claim that subjects actually found these problems to be invalid. But we can now see that they (almost) do, and we might speculate that when yet further violated generic traits (which are nonetheless supposed to be irrelevant to the argument) are also mentioned, subjects will find the arguments as invalid as the ones that they acknowledge as invalid, by applying the second-order default principle of the "rotten egg." This second-order default rule seems to be a clear contender for being added to all normative theories of default reasoning—as well as being a psychologically successful predictor of human judgments.

A central goal of the main study was to investigate whether there are any subtle differences in meaning among three ways of expressing generics. This would form part of the wider study that was started in the pilot study that explored the five different ways mentioned in (5), which in turn is a part of a still wider study that in addition looked at other ways to express generics such as those given in (3). We did discover that there are differences among our three ways, although it remains far from clear what are the underlying semantic features that could possibly give rise to these differences. We discovered that the bare plural formulation contained information that is different from that contained in the generics using a 'most' quantifier or the 'usually' adverb, but that this difference displayed itself only with the simplest of the benchmark problems—the one that asserts the mere existence of some other exception object. So, we discovered that

[BP] Birds fly.
Tweety and Polly are birds.
Polly doesn't fly.
Therefore, Tweety flies.

are judged "more valid," or, "as valid more often," than when the BP premise is replaced by either (Q) 'Most birds fly' or (Adv) 'Birds usually fly.' But in the rather similar benchmark 2, where the only obvious difference is in the presence of an "irrelevant" property of the object-in-question,

[BP] Birds fly.
Tweety and Polly are birds.
Polly doesn't fly.
Tweety is yellow.
Therefore, Tweety flies.

there is no significant difference in subjects' evaluations of the validity of this argument type among the BP, Q, and Adv variants of the generic premise. And in the final problem

[BP] Birds fly.
[BP] Birds eat birdseed.
Tweety and Polly are birds.
Polly doesn't fly.
Tweety doesn't eat birdseed.
Therefore, Tweety flies.

subjects think of the BP version and the Adv version (where both BP premises are replaced by their 'usually' variants) as being the same, and these both as "more valid" than the Q version (where both the BP premises are replaced by their 'most' variants). Just why this should happen with these sorts of problems (which seem so similar to each other) is a question whose answer lies deep in a theory of information and cannot be answered here. But it is clear that there is a difference among the three different generic formulations.

Conclusion

So, are all generics created equal? The data presented here say no—the particular linguistic form in which they are couched will determine subtle differences that show up when the information being put forward by the generic statements are employed in other tasks, such as everyday, default reasoning about the commonsense knowledge that is given by means of these statements. So, the presumption by both the holders of the inductivist view of generics and the holders of the rules-and-regularities view of generics seem to be shown wrong: there is no one interpretation that is correct for all generics.

But is the proposed variant of Carlson's thesis thereby affirmed—that the lexically different formulations of generic statements correlate with the two different sources for information that Carlson identified as forming the basis for different theories of generics? This seems not so clear, since the differences that show up in interpretation of the different linguistic expressions of genericity do not seem to straightforwardly map onto Carlson's inductivist versus rules-and-regularities distinction. And in any case, these results identify *three* different patterns that the generic statements obey, not just the two patterns that the Carlsonian hypothesis suggests. So, before such an identification can be firmly made, a better and deeper theory of the sort of information that is conveyed by generics is needed, and a better and deeper theory of the way information is employed in default reasoning is required. Such theories should set an agenda for researchers involved with the formal (and informal) semantics of linguistic generics and generalizations about the world, and an agenda for researchers involved in the ways people employ this sort of knowledge in everyday reasoning.

Appendix

The BM1.NK.M Problem using the Conifer Story

A conifer is a type of tree or shrub that bears reproductive structures called cones. Among the different conifers are the pines and cedars that grow so commonly in the Vancouver area, as well as throughout the world. But not all conifers are so widespread. In northwestern Saskatchewan there is a conifer commonly called pigwaisa that is found nowhere else, and in the Sonoran desert of northern Mexico is another conifer that is called sumunaro, which is found only in a few other locations around the world. The typical conifer has needle-like structures rather than ordinary leaves, although the sumunaro is an exception to this because it does have medium-sized, ordinary leaves.

According to the information in this paragraph, what kind of leaves do you conclude that pigwaisa has?

The BM2.AK.Q Problem using the Medieval Musical Instrument Story

The middle ages saw a great proliferation of musical instruments, especially stringed instruments. Included in these new stringed instruments were the ganbaz and the tanfir. Most of the new stringed instruments were made of wood, although the gambaz was an exception, being partially metallic and partially animal skin. The tanfir had doubled strings and frets. Both instruments enjoyed wide popularity for more than a century.

What do you think the tanfir was made of, given the evidence presented in this paragraph?

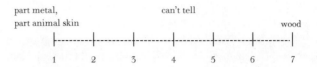

The BM1.AK.Inv (Invalid) Problem using the Pillows Story

Over the years many different styles of pillows have been developed. Not only has the filling been changed from the traditional feather or down, but also there have been many new shapes for pillows. Two new shapes have been the certival support and the lumnaxious pillows. The new shape pillows are often triangular. Some of the certival support pillows have been filled with the traditional goose down.

What do you think the lumnaxious pillows are filled with, according to the evidence presented in this article?

ACKNOWLEDGMENTS

I thank student research assistants Cam Clark and Masha Tkatchouk, and both (Canadian) NSERC grant 5525 and (Simon Fraser University) President's Research Grant for support to them and this research. Also, I offer sincere thanks to the many comments from the conference participants that I received at the 2006 conference, where I presented the results of the pilot study. And thanks to Alasdair Urquhart for technical assistance.

REFERENCES

Carlson, G. (1980). *Reference to Kinds in English*. New York: Garland Press.
Carlson, G. (1982). Generic terms and generic sentences. *Journal of Philosophical Logic* 11, 145–181.
Carlson, G. (1995). Truth-conditions of generic sentences: Two contrasting views. In G. Carlson and F.J. Pelletier (Eds.), *The Generic Book*, pp. 224–237. Chicago: University of Chicago Press.
Carlson, G., and F.J. Pelletier (2002). The average American has 2.3 children. *Journal of Semantics* 19, 73–104.

Elio, R., and F.J. Pelletier (1993). Human benchmarks on AI's benchmark problems. In *Proceedings of the 15th Congress of the Cognitive Science Society*, pp. 406–411. Hillsdale, NJ: Lawrence Erlbaum.

Elio, R., and F.J. Pelletier (1996). On reasoning with default rules and exceptions. In *Proceedings of the 18th Congress of the Cognitive Science Society*, pp. 131–136. Hillsdale, NJ: Lawrence Erlbaum.

Krifka, M., F.J. Pelletier, G. Carlson, A. ter Meulen, G. Chierchia, and G. Link (1995). Genericity: An introduction. In G. Carlson and F.J. Pelletier (Eds.), *The Generic Book*, pp. 1–124. Chicago: University of Chicago Press.

Lakoff, G. (1972). Linguistics and natural logic. In D. Davidson and G. Harman (Eds.), *Semantics of Natural Language*, pp. 232–296. Dordrecht: D. Reidel.

Lakoff, G. (1973). Hedges: A study in meaning criteria and the logic of fuzzy concepts. *Journal of Philosophical Logic 2*, 458–508.

Lawler, J. (1973). Studies in English Generics. Ph.D. thesis, University of Michigan.

Lifschitz, V. (1989). 25 benchmark problems in nonmonotonic reasoning, v. 2.0. In M. Reinfrank, J. de Kleer, and M. Ginsberg (Eds.), *Nonmonotonic Reasoning*, pp. 202–219. Berlin: Springer.

Pelletier, F.J., and R. Elio (2002). Logic and cognition. In P. Gärdenfors, J. Wolenski, and K. Kijainia-Placet (Eds.), *In the Scope of Logic, Methodology, and Philosophy of Science*, Vol. 1, pp. 137–156. Dordrecht: Kluwer.

Pelletier, F.J., and R. Elio (2005). The case for psychologism in default and inheritance reasoning. *Synthèse 146*, 7–35.

Pelletier, F.J., and L. Schubert (2003). Mass expressions. In F. Guenthner and D. Gabbay (Eds.), *Handbook of Philosophical Logic*, 2nd ed., Vol. 10, pp. 249–336. Dordrecht: Kluwer. Updated version of the 1989 version in the 1st edition of *Handbook of Philosophical Logic*.

5

Stability in Concepts and Evaluating the Truth of Generic Statements

JAMES A. HAMPTON

In this chapter I present arguments and evidence in favor of an extended version of prototype theory of concepts (Hampton, 1979, 1995, 1998; Rosch and Mervis, 1975). I first present a brief outline of the theory, and then focus on two particular sources of psychological evidence that support the theory: vagueness in categorization, and the effects of modifiers on the truth of generic sentences.

Prototype Theory

The prototype theory of concepts was introduced by Eleanor Rosch and Carolyn Mervis as an account of a number of phenomena that failed to be explained by classical theories of concepts. The notion of prototype representations had already been proposed in the psychological literature (Dennis et al., 1973; Posner and Keele, 1968), but it was Rosch and Mervis (1975) who provided the most substantial evidence that the concepts underlying the meaning of many substantive terms may be represented in this way. Related work on prototypes in linguistics can be found in Aarts et al. (2004), Lakoff (1987), and Taylor (1995).

In the light of research on prototypes from the last three decades, four main effects can be identified:

a. The membership of conceptual categories is often *vague* in the sense that even though the objects and concepts in question are perfectly familiar, it is difficult to decide whether or not the object belongs in the category. The case of tomatoes being fruit is a classical example, but there are many others (see Hampton, 1979, 1998; McCloskey and Glucksberg, 1978).
b. The different members of a category differ in their *typicality*, even though they may all be clearly in the category. For example, an apple is considered a more typical fruit than a coconut.
c. When people turn their attention to why certain objects fall in a particular category, they are frequently unable to provide a definition that satisfactorily explains the difference between items that are and are not category members (Hampton, 1979; McNamara and Sternberg, 1983). If there is a definition of the term represented in the mind, it appears to be "*opaque*" —there is a sense that one knows the meaning of the term, but it is not transparent how it is applied.
d. When asked to describe characteristics of the category in question, people find it easy to generate many relevant attributes (they may say, e.g., that fruit is sweet or that it grows on trees), but most of these attributes are not in fact true of all members of the category. In fact, they reflect *generic* attributes of the concept.[1]

Put together, these four sources of evidence point strongly to the notion that when people represent a concept such as FRUIT, what they have in mind is some prototypical notion of the typical attributes to be found in the category. What they do *not* have in mind is some rule that tells them what is in the category and what is not.

The four phenomena of vagueness, typicality, opaque definitions, and genericity are readily explained by the central assumption of prototype theory, namely, that people represent concepts in terms of a cluster of attributes that are typically true of the conceptual category.[2]

---

1. There are many forms of generic sentence that have in common that their truth appears to be independent of the existence of counterexamples (see Krifka et al., 1995). Most of the descriptions generated in describing conceptual contents refer to attributes that are frequently true ('birds fly') or relevant ways in which the concept can be distinguished from other contrasting categories ('birds lay eggs').

2. In linguistic semantics, there is evidence for the need to differentiate different domains of concept such as natural kinds, artifacts, and life forms (Wierzbicka, 1984). The psychological literature contains no strong evidence for such a fine grain of distinctions, although there are important differences in how people understand the ontology of natural kinds versus artifacts. In particular, the four phenomena described here are easily demonstrated for a wide range of semantic categories from different domains.

Extended Prototype Theory

Rosch and Mervis (1975) demonstrated the prototype nature of concepts with a simple "feature listing" task in which people had to report the features that they thought might explain category membership. Hampton (1979) adopted a more in-depth interview method in which people were asked to generate attributes of concepts in a number of different ways, but the representation used for each category was again a simple list of features. More recently, evidence has accumulated that the weight that a feature carries in determining the membership of a concept is not just a function of its statistical distribution (as had been originally assumed). Features may be equally strongly associated with a concept class, but differ in the degree to which they are important for membership. For example almost all motor tires are black and round, but it is clearly more likely that a white round object could be a tire than that a black square object could be a tire. In order to explain this important effect, it is now commonly assumed that concept representations also include information about causal dependencies among the features (Barsalou and Hale, 1993; Murphy and Medin, 1985). Being round is causally involved in the function of a motor tire in a way that being black is not. Hence, the representation of concepts includes not only diagnostic information (cues statistically associated with the category) but also explanatory information (why cues co-occur in a particular cluster).

To some writers, this newer theory of concepts leaves prototypes behind. There has been a tendency to see a "prototype" as a visual image corresponding to the most typical individual member of a category (see, e.g., the definition of prototype in Osherson and Smith, 1981), and it is clear that such a representation would be far from adequate in explaining our knowledge of the contents of the concept. If seen as a cluster of properties with associated dependency information, then the extended notion of a prototype is the same notion as a *schema* or *frame* representation. Why, then, still claim that these richer representations are prototypes? Because the four phenomena listed above still need an account, and the most obvious account is that people represent the central tendencies of classes and not their boundaries. The vagueness of categories and the variation in typicality of members are entirely consistent with the schematic representation of the concept including causal dependency information. Similarly, the difficulty in pinning down a precise definition and the ease of producing generic information, regardless of whether or not it is universally true, are both consistent with a representation of a rich prototype concept.

Having outlined the theoretical framework within which the research is set, the remainder of this chapter describes two new lines of research that explore the nature of prototype representations. The first relates to the stability of semantic judgments, and the second, to how one judges the likely truth of generic sentences. Further details of these and related studies may be found in Hampton (2006).

Stability of Semantic Judgments

People have a hard time agreeing about what items belong in what class. Furthermore, they may frequently change their minds. A classic study by McCloskey and Glucksberg (1978) presented students with lists of items related to different categories. For example there were lists of possible furniture, possible fruits, or possible vehicles, including some items that were clearly category members, some that were clearly not, and others that lay somewhere in between.

One group was asked to judge how typical each item was in its category. A second group simply had to judge whether or not the item belonged in the category, with a yes/no binary decision. This second group returned a few weeks later and made the judgment for a second time. Figure 5.1 illustrates their results. The horizontal axis shows intervals of mean typicality derived from the first group, and the vertical axis shows the mean probability that the second group categorized the item as a category member. Note how the probability of saying "yes" rises in a very systematic fashion as typicality increases. Notice also how for a wide range of items the probability of categorization is neither 1 nor 0. This region of the borderline of a category

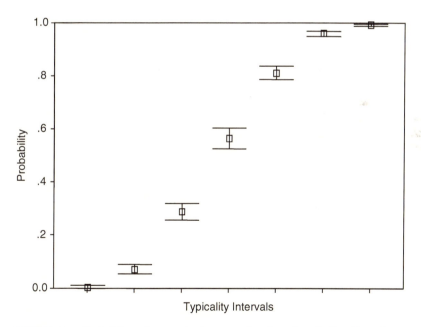

FIGURE 5.1 Probability of categorization as a function of typicality (data from McCloskey and Glucksberg, 1978)

is where the phenomenon of vagueness is to be found. Items within this region are neither clearly in the category nor clearly not in it, at least on the basis of the lack of agreement among the students. Furthermore, McCloskey and Glucksberg (1978) demonstrated that for these same borderline items people in the second group were likely to change their categorization from one week to the next. The vagueness, therefore, was not simply a matter of different strokes for different folks—many individuals were unable to maintain a consistent decision about the categorization at the borderline.

In relation to the question of stability, three studies are presented here. The first concerns the question of what underlies the instability. People were asked to choose between different justifications of why they chose to say that an item was neither clearly in a category nor clearly out. The second study considered whether the region within which instability occurs might itself be well defined. Are people clear about what is "clearly" in or out of a category and what is not? The final study concerned the stability of typicality judgments and, in particular, whether they follow a prediction of prototype theory that typical items should show greater stability than atypical items.

Study 1: The Reasons for Vagueness

A number of different theoretical accounts have been offered to explain the problem of vagueness in categorization (see Hampton, 2007). In this study, I focused on three possibilities:

a. People are vague about categorization because they feel that they do not know enough about the concept or the item being categorized. This view is known as the epistemological account of vagueness (Williamson, 1994), by which there is actually a correct meaning for a term, but people are *ignorant* of it.
b. People are vague about categorization because the terms are inherently *ambiguous*, and the precise answer would require more information about the exact context involved.
c. People are vague about categorization because the concept is itself vague —there just is no hard-and-fast way of determining the "correct" categorization status of an item, even if one knew all about it and even if the context was fully specified. One way that this can manifest is if membership in a category is graded or *partial*, so that whether one counts an item as belonging or not would depend on whether a broad or a narrow view of the category was taken.[3]

---

3. There is also the possibility that the contents of the concept are in some sense "indeterminate"—a person's knowledge of the term is "woolly," and they possess no recognizable set of descriptors that would enable it to be applied in any consistent way—perhaps because no such descriptors exist. Since the experimental materials here

In order to test these three ideas, students were given lists of items to categorize, including many that were borderline. They were given three responses to choose from: "clearly yes," "intermediate," and "clearly not" and worked through the lists, categorizing each item with one of these responses. When they reached the end, they were unexpectedly asked to revisit each of the responses where they had marked the categorization as "intermediate" and tick off one or more possible explanations from a list provided. The list included two possible types of ignorance (e.g., "because I don't know enough about the category/item to say"), four reasons that related to ambiguity (e.g., "because some examples of the item are in the category and some aren't" or "because it depends on the context in which you have to categorize the item"), and one that related to partiality ("because it depends on whether you take the category in a broad or in a narrow sense"). Table 5.1 shows the options given to the participants together with the percentage of decisions for which they were chosen, and table 5.2 gives the relative percentages divided into the three main types of reason for each of the eight categories. Ignorance, ambiguity, and partiality were each selected about one-third of the time as explanations for an intermediate response, but the frequencies varied across categories.

There was a clear differentiation between the categories that may be considered to have some kind of technical definition (the biological categories, vegetable, fruit, fish, and insect, and the science category) and the rest. Ignorance was used as an explanation 49% of the time for these five categories, and only 8% of the time for the rest (the sports, tools, and furniture categories). Table 5.1 confirms that ignorance largely related to particular items, rather than to the category as a whole. This result is

TABLE 5.1 Reasons offered to participants in study 1 for choosing an "intermediate" categorization response

Reason	Percentage
A. because I don't know enough about the category to say	7%
B. because I don't know enough about the item to say	25%
C. because some examples of the item are in the category, and some aren't	3%
D. because it depends on how you define the category (otherwise than G)	15%
E. because it depends on how you define the item	11%
F. because it depends on the context in which you have to categorize the item	7%
G. because it depends on whether you take the category in a broad or in a narrow sense	31%
H. some other reason (please specify at the bottom of the page)	1%

A and B were classed as ignorance, C–F were classed as ambiguity, and G was classed as partiality. Data are percent usage across items and participants.

employed well-known familiar concepts, I do not consider this explanation for vagueness any further here.

TABLE 5.2 Percentage of reasons given, divided into three main types of reason and individual category

Category	Type of Reason		
	Ambiguity	Partiality	Ignorance
Vegetable	15%	23%	62%
Fruit	27%	13%	60%
Fish	30%	33%	37%
Insect	26%	20%	53%
Science	37%	29%	35%
Sport	49%	43%	8%
Tool	55%	35%	9%
Furniture	50%	41%	8%
Mean	29%	31%	32%

consistent with a finding reported by Estes (2004) that people are inclined to assume that membership of natural kind categories is all-or-none, while membership in artifact categories may be partial. In line with Estes's results, ambiguity and partiality, by contrast, were more commonly selected in the latter three categories. Ambiguity was selected 51% of the time for the nonnatural kinds and 27% of the time for the natural kinds, while partiality was selected 40% and 24% of the time, respectively.

A final analysis looked at the correlation across items of the frequency with which different justifications were selected. The two ignorance options were uncorrelated with the rest, whereas the different forms of ambiguity and partiality all correlated positively. The degree to which people stated that there was ambiguity in the item, category, or context correlated with the degree to which people stated that the answer depended on whether the category was taken in a broad or a narrow sense. There was therefore evidence in the correlations that there are principally two underlying reasons for people being unsure of a categorization—one relating to ignorance and the other relating to the fact that the concepts are vague and context dependent so there is no determinate answer to the question. The latter explanation clearly fits well with a prototype account of concepts. Because we represent concepts via their central cases, and do not hold consistent information about the rules for determining the boundaries, we are often aware that membership in a category can be a matter of how broadly or narrowly one draws the criterion for category membership. Naturally, this study does not provide definitive evidence of the reasons that people may *actually* have for giving an intermediate judgment—we are relying on their ability to introspect about their concepts and their conceptual judgments. However, this kind of metalinguistic judgment has proved of considerable value in other fields, and to the extent that the reasons chosen differ systematically between items and domains they provide an additional source of behavioral evidence that a psychological account of vagueness needs to address.

Study 2: Second-Order Vagueness

A second question that arises about the vagueness of semantic categories is whether, in fact, the borderline region in which categorization is uncertain is itself more tightly defined. Accounts of the logic of vagueness have been offered that use three-valued logics according to which a statement can be true or false or have an undefined truth status (Halldén, 1949; Williamson, 1994). It is obviously important for such logics that the question of whether or not a statement has a truth value should not itself be subject to vagueness. A recent example of this type of three-valued account is Kamp and Partee's (1995) use of "supervaluations" to explain the logic of vague statements (Fine, 1975; Kamp, 1975). They hypothesized that the category scale can be divided into three distinct regions—clear members, partial members, and clear nonmembers. They then offered their supervaluationist account of how one can still deduce that 'x is an A or not an A' is true, and that 'x is an A and not an A' is false, even when the question of whether x is an A or not is itself undetermined (because x is borderline to the category A). Their account thus aims to explain how I may be stumped by the question of whether a person living in France with assets of $100,000 should be categorized as "rich" but would be happy to agree that they are definitely either rich or not rich and that they are definitely not both rich and not rich. In order for this account to work, however, the borderline indeterminate region of membership needs to be clearly demarcated. The second study examined whether this was the case with borderline cases of category membership.

The data collection was very simple. Two groups of students were used, and each returned after two weeks to repeat exactly the same task. The degree to which they changed their responses on the second occasion was taken as a measure of the vagueness of the decision. One group was given a standard yes/no categorization to make about a list of cases like those used by McCloskey and Glucksberg (1978). The second group was allowed to make one of three responses, corresponding to the Kamp and Partee (1995) three-valued logic. Thus, they could say "definitely yes," "maybe," or "definitely no" to each category item. If it is the case that we clearly recognize when we know something to be true or false for sure, then one would expect a lower rate of inconsistency for the second group when they returned to make their judgments a second time. Intuitively, it seems quite plausible that I could know an apple to be a fruit and a carrot not to be a fruit, and I know equally well that a tomato is unclear. So two weeks later I shouldn't be likely to change my mind. On the other hand, if I can only say "yes" or "no," I may say yes to the tomato the first time around and then change it to a no on the second occasion, since I really am unclear about the correct answer. A well-delimited region of uncertainty should therefore correspond to a reduced level of inconsistency in the three-response group compared with the two-response group.[4]

The study has now been run twice in collaboration with students at City University London, once with Bayo Aina, and once with Gurinder Jai. On the replication study, we tightened up the instructions and asked people in the second group to say only "yes" or "no" if they were sure in their minds about the answer. The results were quite clear and consistent. There was no observable change in either study in the rate of inconsistency between the two groups. In order to compensate for the fact that the second group had greater opportunity for change (having three rather two response options), the consistency was first counted for both groups as the probability of a "yes" or a "clear yes" being unchanged on the second occasion. The same calculation was then made for a " no" or a "clear no," and the average of yes and no stability was calculated for each group.

In Aina's experiment, with 58 participants and a list of 132 items, the percentage of consistent responses was $83.2 \pm 1.8\%$ for the two-response group and $83.7 \pm 1.5\%$ for the three-response group.[5] The difference (0.5%) was clearly not significant, and the design had an estimated power of 90% to detect a significant difference between the participant groups of as little as 4%, with alpha set at .05.[6]

In the smaller replication by Jai (with 40 participants and 96 items), the two-option group showed average stability of yes and no responses of 67%, while the three-option group showed average stability for clear-yes and clear-no responses of 69%. Again, there was no significant difference. This study had an 80% power to detect a difference of 10% by subjects and 5% by items, with alpha at .05.

There was therefore no evidence that people have any better idea about whether an item is vague than they do about whether an item is in the category in the first place. The result is entirely consistent with the view of concepts that has been proposed here. We represent central cases rather than rules for categorization. Borderline cases are not all just "known exceptions" or problem cases—an item that seems borderline on one occasion may appear quite clear on the next occasion. In fact, in the Jai study an

---

4. A reviewer correctly pointed out that inconsistency across time is not incompatible with Kamp and Partee's supervaluationism, since the determination of the vague region only has to be held constant between the evaluation of the proposition 'x is an A' and the proposition 'x is not an A.' However, the question of whether we "know what we don't know" about categorization remains an important and interesting empirical question. The existence of fairly precise borders for the region of vagueness would place a useful constraint on theoretical accounts of instability and vagueness in categorization.

5. The $\pm$ figures reflect 95% confidence intervals.

6. The analysis by items had a 90% power to detect a difference as small as 2.2%.

item identified as borderline on the first occasion had only a 20% chance of being still considered borderline at retest.

Study 3: Stability of Typicality Judgments

The last of the studies of stability in concepts was an experiment that tested a direct prediction of prototype theory concerning typicality judgments. The typicality of an item is usually measured by asking people to rate how "representative" or "typical" it is of a particular category. People commonly agree, for example, that a chair is a typical kind of furniture whereas a piano is atypical. The account offered of this phenomenon (Rosch, 1975) is that people find it quite easy to compare an item with the prototype representation of the category and so to judge how similar the two concepts are. Indeed, it is assumed that judging typicality and judging category membership involve similar processes—assessing the degree of match between the item and the category—although there is evidence that the weight accorded to deeper features may be greater in the case of categorization (Hampton, 1998; Rips, 1989).

The experiment tested the simple notion that variability in typicality judgments (which Barsalou [1987] showed to be quite considerable) could be explained by variability in the weight of different attributes in the concept representation. (This same variability may also be a primary source of the instability in categorization reported above.) If this were the case, then one would predict that the closer the match of an item to a category, the *less* variable the rating of typicality should be. For example, an item that matches *all* of the attributes of the category will always have maximum similarity to the concept, regardless of attribute weights. The typicality judgment should therefore always be maximal. But an item that matches, say, two-thirds of the attributes will be subject to instability depending on whether the two-thirds of the attributes that it has are given high or low weight on a particular occasion. A similar stability should also be found for items that match very few attributes—but in that case they would clearly not be category members, and so questions of typicality would not arise.

The experiment was again quite simple. A single group of students rated typicality of lists of items in a number of categories on two occasions two weeks apart. Items were chosen to fall at equal intervals on a scale of mean typicality obtained in pretesting. The likelihood of the same rating being given again was calculated as a function of the initial rating, and is shown in figure 5.2. There was clearly greater stability at *both* ends of the scale—as reported by Barsalou (1987). However, this can easily be explained by an end anchor effect. If, for example, an item was considered very atypical initially and given a rating of 9, then on the second occasion it could be seen either as more or as less typical than before. If more typical, then the rating would improve to 6 or 7, but if less typical the rating would remain at 9. The end points should therefore show less instability than mid points on the scale for this reason. Our prediction, however, related to the difference between

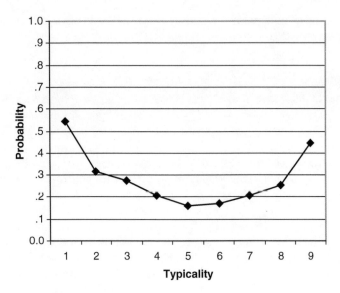

FIGURE 5.2 Probability of same rating

typical and atypical items, and this difference is also seen in figure 5.2. The mean stability for the high typicality end of the scale (ratings 1–4) was significantly greater than that for the low typicality end of the scale (ratings 6–9), in keeping with the prediction. Being close to the center of a category appears to provide greater stability to the judgment of similarity to the prototype.

Conclusions from the Stability Experiments

Three studies have been presented investigating how and why semantic judgments tend to be unstable. The first study showed that when people said that a categorization was uncertain, they then tended to choose a variety of explanations. In keeping with the notion that people represent the center of categories rather than their borders, people often considered that uncertainty arose because conceptual categories can be interpreted in a broad or a narrow sense, and can be ambiguous and context dependent. The second study showed that borderline cases of categories are not a separately identifiable class of item—when people have to judge whether or not a case is borderline, they are just as likely to change their response as when they have to judge whether or not it is in the category. The borderlines are just fluid and vague, and three-valued logics cannot be readily applied to categorization. The final study showed that, as predicted by prototype theory, typicality judgments are more stable for high typical items than for low typical items. In the second part of this chapter, I turn to an account of some experiments on a second phenomenon associated with prototype theory—genericity.

Genericity: Why Do People Say That Birds Can Fly?

Study 4: The Modifier Effect

Genericity is the aspect of prototypes that relates to the attributes that people produce when describing members of a conceptual category. When asked to say why something is a bird, or what makes an activity a sport, people typically generate "features" or attributes that are *typically* true of the category. Not only do different things belong to categories to different extents, but attributes may also be true of categories to different degrees. When we say 'birds fly' or 'sports provide exercise, 'we have in mind that these are important parts of the information that anyone who possesses these concepts should know. They are part of the general culturally moderated cluster of knowledge that is tied to these particular concepts. There may well be more scientific ways in which conceptual categories can be defined, particularly for natural kind terms such as bird or gold, but the average person's notion may be little more than this cluster of interrelated attributes. There is some debate about whether people also subscribe to the notion of a scientifically discoverable "essence" that constitutes the real meaning of natural kind terms. Hampton et al. (2007) found evidence that there may in fact be individual differences in how willing people are to make this assumption.

The studies of genericity to be discussed form part of a series of experiments reported by Jönsson and Hampton (2006, 2007), who followed up an interesting result discovered by Connolly et al. (2007), what we called the "modifier effect." It turns out that if one group of people are asked to judge how true it is that, say, ravens are black, and another group is asked to judge how true it is that young jungle ravens are black, then the second group will give lower ratings. The modifier phrase 'young jungle' has an effect of reducing the judged truth of a property of the head noun 'ravens.' This effect was replicated by Jönsson and Hampton (2007), who showed (as did Connolly et al., 2007) that the effect is moderated by the typicality of the modifier. That is to say, there is a small effect for a typical modifier ('feathered ravens are black') but a larger effect for an atypical modifier ('jungle ravens are black').

The theoretical significance of this effect relates to what it has to say about the process of combining (or modifying) concepts. According to Connolly et al. (2007), the modifier effect shows that the prototype of a modified concept (e.g., jungle raven) bears no direct relation to that of its unmodified parent concept (raven). They argued that to assume that some atypical subset of ravens would have the same stereotypical properties as the general class would be a mistake, and so they claimed that the reduced truth for modified sentences shows that people do not usually make this mistake. According to the theory of conceptual combination that they espoused, concepts such as JUNGLE and RAVEN combine first through intersection of their referential sets (presumably the class or species of ravens that are found in or adapted to living in jungles). All the possessor of the concept is then allowed to assert with any confidence is that jungle ravens are ravens and that they bear some important relation to jungles. They cannot be assumed to be black with the same confidence with which it is known that ravens are black. This theoretical position is opposed to one commonly adopted in cognitive science that suggests that there are a set of processes by which the prototypes of two concepts may be combined to yield a composite (e.g., Cohen and Murphy, 1984; Hampton, 1987, 1988; Murphy, 1988; Smith et al., 1988; Wisniewski, 1997). According to prototype combination models, modified concepts should indeed "inherit" stereotypical properties from their "parent" concepts. Hampton (1987) provided detailed data on precisely how the degree to which attributes are judged true of modified concepts can be derived from their judged truth for the simple concepts from which they are combined.

In our first experiment (Jönsson and Hampton, 2007, experiment 1), we showed that the effect seems to be one that generates a systematic reduction in the truth of properties. There was a significant correlation such that the more true a statement was rated when unmodified, then the more true it was also rated when modified. In a second experiment, we showed that there are possibly three different sources of the effect. We gave students pairs of sentences, one modified and one unmodified, and first asked them to judge which (if either) was more true. About 60% of the time they claimed that the two sentences were equally true, supporting the idea that the modification had been incorporated into the concept prototype with little additional effect to the other prototypical properties. On the remaining occasions, the modifier effect was seen—most often the modified sentence was judged less likely. We then asked our participants to tell us the reason that they had made this choice, and obtained three main types of justification.[7] The first, and most common, was simply to say that the modified sentences did not sound sensible, because the property was in any case true of the whole class. For example, one participant preferred to say 'seaweed is green' rather than 'Indian seaweed is green' because seaweed (in her view) was all green anyway. The modified sentence violates Grice's maxim of quantity ("Be as informative as you can" see Grice, 1975; Levinson, 2000). The second most common justification depended on the participant thinking of a way in which the modifier might affect the property directly or indirectly. The modifiers had been chosen to be independent of the properties, but participants often came up with unanticipated connections. For example, it was considered less likely that bitter nectarines would be juicy

---

7. As before, we are assuming that participants have relevant intuitions about the basis of their judgments that should be used to constrain theoretical explanations.

than that nectarines would be juicy, on the grounds that the bitterness might signal unripeness, and unripeness would lead to less juice.

The two types of justification found so far do not really provide any reason for supposing that people did not consider that (in the absence of any knowledge-based reasoning) modified concepts inherit the properties of their parts. The third type of justification, however, did support Connolly et al.'s (2007) position to a degree. This class of reasons related to the uncertainty that an unknown subclass would have the stereotypical properties of the general class. Thus, for example, one participant did not think that Brazilian doves were as likely to be white as doves in general, because "I don't know any Brazilian doves, so they may be any color." The uncertainty of properties of atypically modified concepts was clearly insufficient as a general account of the modifier effect, however, since it accounted for less than 10% of the judgments that participants made in the task.

A further experiment in Jönsson and Hampton (2007) showed another connection between the properties of a concept and those of an atypical subset of the concept. We manipulated the degree to which properties were considered immutable or mutable for the concept. The notion of mutability of a property of a concept (Sloman et al., 1998) is that properties differ in terms of how easy it is to imagine the concept, otherwise unchanged, as lacking that property. It is assumed that mutability reflects the degree to which a property is seen as causally linked to other properties. Thus, it is easy to imagine a tire that is not black or made of rubber, but harder to imagine a tire that is not round or with no means of attaching it to a wheel or axle. In our experiment, we showed that mutability affects the degree to which a property is considered true for modified versions of a concept. Thus, a property that was more mutable for the unmodified concept ('ravens are black' as opposed to 'ravens have feathers') was considered less likely to be true of the modified concept. Hence, 'young jungle ravens have feathers' was considered more likely to be true than 'young jungle ravens are black,' because of the lower mutability of the former property for ravens.

Study 5: The Inverse Conjunction Fallacy

If prototype theory is right, then people should find it difficult to make consistent decisions about whether one class is completely included within another. For example, if one decides that chairs are a kind of furniture on the basis of the match between one's representation of the concept of chair and one's representation of furniture, what is to make sure that there aren't some atypical chairs that are not actually furniture? Focusing on the descriptive content of the central cases of a category makes it hard to determine what the extent of the denotation of the concept should be, and harder still to determine whether a strict relation of class inclusion holds between two concepts. Hampton (1982) demonstrated this effect with artifact categories. In that study, I showed that people were quite happy to say that a chair is a kind of furniture, even though they judged car seats or ski lifts to be chairs that are not furniture.[8] In Hampton (1988, experiment 1), I similarly showed that the class of things called "school furniture" or "office furniture" was not fully contained in the class of things called "furniture."

As applied to the modifier effect discussed above, these earlier results suggested that we might find that people continue to consider a modified sentence less likely to be true, even when it is universally quantified. Given the generic semantics of a sentence such as 'ravens are black,' it is quite permissible for some unusual kinds of raven to be white without thereby rendering the statement false. 'Ravens are black' says something about what people generally believe is true of typical ravens (it expresses a prototypical property of ravens), but it does not say that *all* ravens are black. On the other hand, the statement 'All ravens are black' should logically entail that there are *no* kinds of raven that are not black. So if one judged this sentence true, one should also logically judge to be true all sentences of the form 'All M ravens are black' where M is some modifier that selects some nonempty subset of ravens.

Jönsson and Hampton (2006) gave students a choice of two sentences of the following form and asked which was more true (or were they equally true):

All sofas have backrests

All uncomfortable handmade sofas have backrests

There was a strong tendency for people to judge that the unmodified sentence was more likely to be true than the modified sentence, regardless of the illogicality of this position. (After all, "all sofas" should include all sofas, even those that are uncomfortable and handmade.) This effect, which we termed the inverse conjunction fallacy, is similar in many respects to Tversky and Kahneman's (1983) well-known conjunction fallacy. The conjunction fallacy involved the belief that a person was more likely to be found in a subset of a class than in the class itself. For example, a woman (Linda) with liberal politics was judged more likely to be a feminist bank teller than to be a bank teller. The inverse conjunction fallacy turns the effect upside down by considering whether a *property* is true of a whole class or a whole subclass, and in this case the fallacy is to believe that it is more likely for the property to be true of the whole class than of the subclass.

---

8. It may be objected that although people may categorize a car seat as a chair, they would never call it such in everyday discourse. Different accounts may therefore be needed to explain judgments of concept subordination and naming behavior. However, both are equally relevant to the question of how our concepts are represented. Fortunately, Hampton's (1982) intransitivity result did not depend on this single case, but was also found with other kinds of furniture (clocks, shelves, mirrors, lamps, etc.) and other general categories (sports, pets, plants, tools, etc.) where the objects (e.g., Big Ben) could also be named by the concept term (e.g., clock).

The explanation for both kinds of conjunction fallacy may be the same. Tversky and Kahneman (1983) referred to something they called "intensional reasoning"—that is, judging the likelihood of an event or fact on the basis of similarity or representativeness.[9] Because Linda is more typical of a feminist than of a bank teller, it somehow seems more likely that she should be in the conjunction of the two sets than in just one of them alone. The inverse conjunction fallacy also seems to involve intensional reasoning. Whatever led to the modifier effect with generic sentences also seems to apply to universally quantified sentences.

Jönsson and Hampton (2006) went on to investigate the fallacy further. First they showed that it does not depend on use of the word "all." Very similar fallacious responding was found with the phrases 'All X are Y,' 'All X are always Y.' '100% of X are Y,' and 'Every single X is Y.' In other words, no matter how the meaning of the universal quantifier was expressed, it tended to be ignored. For example, people judged it more likely that 'every single sofa has a backrest' than that 'every single uncomfortable handmade sofa has a backrest.'

A second control was to check whether people believed that the subclass was in fact in the class. A group of students judged sentences such as 'All jungle ravens are ravens' or 'All uncomfortable handmade sofas are sofas.' For the large majority of these sentences, all judges agreed that they were true, so the effect was not owing to doubt about class membership. A third control looked at whether people thought that some subsets might not exist. If there are no such things as jungle ravens, then one would be justified to say it is unlikely to be true that all jungle ravens are black (in the sense that it would not be true because it would be meaningless). A few of the modified concepts turned out to generate some doubt as to their existence, but there was no evidence that these concepts were any more likely to generate the fallacy than were others.

In our final experiment, we aimed to discover conditions under which the fallacy would be reduced. Placing the sentences side by side and asking which (if either) was more true did not change the incidence of the fallacy compared with having different groups of students judge each separately. However, when people were given the two sentences as a pair, one after the other, and simply had to say whether each one in turn was true or false, the incidence of the fallacy was significantly diminished. Committing the fallacy therefore is not inevitable, but it appears to be a very powerful cognitive illusion, resisting most of our efforts to alleviate it.

---

9. Intensional reasoning more generally will depend on what one assumes to be the "intension" of a concept. Tversky and Kahneman's (1983) account clearly presupposes that people make judgments of likelihood on the basis of how similar an object or situation is to the prototypical case. For them, an intension thus includes some kind of prototype.

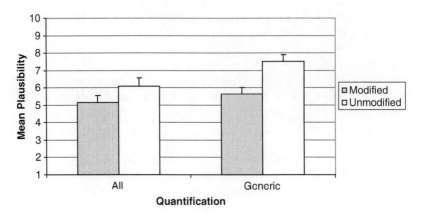

FIGURE 5.3 Effects of modifiers on plausibility

Study 6: Comparing the Modifier Effect and the Inverse Conjunction Fallacy

The final study described here is an unpublished experiment conducted with Ou Lan, a student at City University London (also reported in Jönsson and Hampton, 2007). In this experiment we considered what effect the introduction of universal quantifiers would have on the modifier effect. Does the word "all" simply reduce general confidence in the statements, in a similar way to the "atmosphere effect" described for syllogisms by Woodworth and Sells (1935)? Or would the difference between modified and unmodified sentences be reduced when the logical constraint of "all" is introduced. The experiment involved two factors. Sentences could be modified or unmodified as before, and in addition they could be generic or universally quantified. Figure 5.3 shows the results of the experiment in which students judged the likely truth of sentences of each of the four types. The statistical analysis confirmed significant main effects and a significant interaction. Modified sentences were judged less likely to be true, both when generic and when universally quantified. Introduction of the quantifier reduced the judged likelihood of both modified and unmodified sentences. The interaction reflected the fact that the modifier had a larger effect on the generic sentences than on the universally quantified sentences. Universal quantifiers made people more cautious about judging a sentence likely to be true, and also diluted the effect of the modifier. People still, however, thought it more likely that 'All X are Y' than that 'All M X are Y.'

Conclusions

In this chapter I have endeavored to show how the way in which people use concepts follows systematic patterns that do not necessarily conform to the strict requirements of logic. Human thinking can be classed as logical, illogical, and nonlogical (Evans, 1982). The first two kinds of thinking are correct or incorrect attempts to derive inferences based on the logic of a problem, whereas the last is the use of a process that is neither logical nor illogical—it is a form of rationality that involves approximate, heuristic reasoning. The kinds of results obtained in the studies described here fall best into the latter class of thinking. People need words in order to communicate, and they frequently are not completely clear about what they are trying to say. Communication only needs to be as precise as the context requires. Most situations involve typical cases in which the names of things are clear, and the categories they belong to are uncontroversial. It is usually more important to know what is generally true of a type of thing, than to know whether something is universally true—more useful to know that toadstools are generally poisonous than that toadstools are not all poisonous. We learn the generality of things first, and only with sufficient expertise would we expect to know of detailed exceptions.

The effects of this way of building and representing knowledge are that semantic judgments can often be unstable or vague—we do not know (or necessarily care) whether certain things belong in certain categories, because what we represent are the clear cases. Unclear cases have to be decided by comparison with clear cases in the context of the problem to be solved. Similarly, a judgment of whether or not a property is true of a class may be made on the basis of how similar the class is to other classes for which the property is known to be generally true. We find it hard to differentiate our intuitions about what is generally true and what is always true. That is a sign of thinking intensionally rather than extensionally.

REFERENCES

Aarts, B., Denison, D., Keizer, E., and Popova, G. (2004). *Fuzzy Grammar: A Reader*. Oxford: Oxford University Press.

Barsalou, L.W. (1987). The instability of graded structure: implications for the nature of concepts. In U. Neisser (Ed.), *Concepts and Conceptual Development: Ecological and Intellectual Factors in Categorization* (pp. 101–140). Cambridge: Cambridge University Press.

Barsalou, L.W., and Hale, C.R. (1993). Components of conceptual representation: from feature lists to recursive frames. In I. van Mechelen, J.A. Hampton, R.S. Michalski, and P. Theuns (Eds.), *Categories and Concepts: Theoretical Views and Inductive Data Analysis* (pp. 97–144). London: Academic Press.

Cohen, B., and Murphy, G.L. (1984). Models of concepts. *Cognitive Science*, 8, 27–58.

Connolly, A.C, Fodor, J.A., Gleitman, L.R., and Gleitman, H. (2007). Why stereotypes don't even make good defaults. *Cognition*, 103, 1–22.

Dennis, I., Hampton, J.A., and Lea, S.E.G. (1973). New problem in concept formation. *Nature*, 243, 101–102.

Estes, Z. (2004). Confidence and gradedness in semantic categorization: Definitely somewhat artifactual, maybe absolutely natural. *Psychonomic Bulletin and Review, 11*, 1041–1047.

Evans, J.St.B.T. (1982). *The Psychology of Deductive Reasoning*. London: Routledge and Kegan Paul.

Fine, K. (1975). Vagueness, truth and logic. *Synthese, 30*, 265–300.

Grice, H.P. (1975). Logic and conversation. In P. Cole and J.L. Morgan (Eds.), *Syntax and Semantics* (3rd ed., pp. 41–58). New York: Academic Press.

Halldén, S. (1949). *The Logic of Nonsense*. Upsala: Universitets Arsskrift.

Hampton, J.A. (1979). Polymorphous concepts in semantic memory. *Journal of Verbal Learning and Verbal Behavior, 18*, 441–461.

Hampton, J.A. (1982). A demonstration of intransitivity in natural categories. *Cognition, 12*, 151–164.

Hampton, J.A. (1987). Inheritance of attributes in natural concept conjunctions. *Memory and Cognition, 15*, 55–71.

Hampton, J.A. (1988). Overextension of conjunctive concepts: Evidence for a unitary model of concept typicality and class inclusion. *Journal of Experimental Psychology: Learning, Memory, and Cognition, 14*, 12–32.

Hampton, J.A. (1995). Testing prototype theory of concepts. *Journal of Memory and Language, 34*, 686–708.

Hampton, J.A. (1998). Similarity-based categorization and fuzziness of natural categories. *Cognition, 65*, 137–165.

Hampton, J.A. (2006). Concepts as prototypes. *The Psychology of Learning and Motivation: Advances in Research and Theory, 46*, 79–113.

Hampton, J.A. (2007). Typicality, graded membership and vagueness. *Cognitive Science, 31*, 355–383.

Hampton, J.A., Estes, Z., and Simmons, S. (2007) Metamorphosis: Essence, appearance and behavior in the categorization of natural kinds. *Memory and Cognition, 35*, 1785–1800.

Jönsson, M.L., and Hampton, J.A. (2006). The inverse conjunction fallacy. *Journal of Memory and Language, 55*, 317–334.

Jönsson, M.L., and Hampton, J.A. (2007). The modifier effect in within-category induction: On prototypes and default inheritance. Unpublished manuscript, City University London, September.

Kamp, H. (1975). Two theories about adjectives. In E.L. Keenan (Ed.), *Formal Semantics of Natural Language* (pp. 123–155). Cambridge: Cambridge University Press.

Kamp, H., and Partee, B. (1995). Prototype theory and compositionality. *Cognition, 57*, 129–191.

Krifka, M., Pelletier, F.J., Carlson, G.N., ter Meulen, A., Chierchia, G., and Link, G. (1995). Genericity: An introduction. In G.N. Carlson and F.J. Pelletier (Eds.), *The Generic Book* (pp. 1–124). Chicago: University of Chicago Press.

Lakoff, G. (1987). *Women, Fire and Dangerous Things*. Chicago: University of Chicago Press.

Levinson, S. (2000). *Presumptive Meanings: The Theory of Generalized Conversational Implicature*. Cambridge, MA: MIT Press.

McCloskey, M., and Glucksberg, S. (1978). Natural categories: Well-defined or fuzzy sets? *Memory and Cognition, 6*, 462–472.

McNamara, T.P., and Sternberg, R.J. (1983). Mental models of word meaning. *Journal of Verbal Learning and Verbal Behavior*, 22, 449–474.

Murphy, G.L. (1988). Comprehending Complex Concepts. *Cognitive Science*, 12, 529–562.

Murphy, G.L., and Medin, D.L. (1985). The role of theories in conceptual coherence. *Psychological Review*, 92, 289–316.

Osherson, D.N., and Smith, E.E. (1981). On the adequacy of prototype theory as a theory of concepts. *Cognition*, 11, 35–58.

Posner, M.I., and Keele, S.W. (1968). On the genesis of abstract ideas. *Journal of Experimental Psychology*, 77, 353–363.

Rips, L.J. (1989). Similarity, typicality and categorization. In S.Vosniadou and A. Ortony (Eds.), *Similarity and Analogical Reasoning* (pp. 21–59). Cambridge: Cambridge University Press.

Rosch, E.R. (1975). Cognitive representations of semantic categories. *Journal of Experimental Psychology: General*, 104, 192–232.

Rosch, E.R., and Mervis, C.B. (1975). Family resemblances: studies in the internal structure of categories. *Cognitive Psychology*, 7, 573–605.

Sloman, S.A., Love, B.C., and Ahn, W.K. (1998). Feature centrality and conceptual coherence. *Cognitive Science*, 22, 189–228.

Smith, E.E., Osherson, D.N., Rips, L.J., and Keane, M. (1988). Combining prototypes: A selective modification model. *Cognitive Science*, 12, 485–527.

Taylor, J.R. (1995). *Linguistic Categorization: Prototypes in Linguistic Theory* (2nd ed.). Oxford: Clarendon.

Tversky, A., and Kahneman, D. (1983). Extensional versus intuitive reasoning: The conjunction fallacy in probability judgment. *Psychological Review*, 90, 293–315.

Williamson, T. (1994). *Vagueness*. London: Routledge.

Wierzbicka, A. (1984). Apples are not a kind of fruit: The semantics of human categorization. *American Ethnologist*, 11, 313–328.

Wisniewski, E.J. (1997). When concepts combine. *Psychonomic Bulletin and Review*, 4, 167–183.

Woodworth, R., and Sells, S. (1935). An atmosphere effect in syllogistic reasoning. *Journal of Experimental Psychology*, 18, 451–460.

6

Generics as a Window onto Young Children's Concepts

SUSAN A. GELMAN

"Did you know when a pig gets to be big, they're called hogs?" (mother speaking to three-year-old child)

"Bats are one of those animals that is awake all night." (mother speaking to three-year-old child)

"Adams don't have to take naps." (Adam M., two and a half years old)

The sentences above are perfectly ordinary, even typical of the sort of expressions one would hear if eavesdropping on the conversations between a preschool-age child and his or her parent. Yet these ordinary expressions are anything but mundane in their implications for children's thought. All three sentences include generic noun phrases (NPs; e.g., a pig, bats, Adams), which are distinctive in their reference to kinds versus individuals ("Adams" as a group versus a particular "Adam"; Carlson and Pelletier, 1995; Prasada, 2000). Generic nouns are semantically interesting in how they extend beyond the immediacy of a child's everyday experience. (There is no actual set of "Adams" in the child's experience.) They are formally interesting in the varieties of expression they encompass. And they are conceptually interesting in the information they convey about intuitive theories of the world.

This chapter focuses on the implications of generics for young children's concepts. The heart of the argument is that generics pose a challenging learning puzzle, yet children acquire them with ease. This fact alone suggests that children must possess, at an early age, the conceptual prerequisites to construct and learn from generic language. This argument owes much to prior work on semantic development, which notes that word-learning more generally presents a classic problem of induction (Markman, 1989; Quine, 1960). In brief, we ask how young children acquire generic nouns in the face of ambiguous and insufficient evidence.

This chapter is organized as follows. After considering the learning puzzle that generics present, we review some of the recent evidence that young children start using generics by about the age of two and a half. We then briefly consider possible learning models and what they imply about children's early conceptual system. Finally, we note that children's capacity to understand generics suggests that language input may shape children's concepts and knowledge systems, and sketch some preliminary evidence for such effects. The chapter ends by raising some larger questions that these issues suggest.

The Puzzle of Induction

Researchers studying early semantic development have noted that learning a single word requires an inductive leap well beyond the evidence. When hearing, "This is an *aye-aye*," the child cannot know a priori whether the word "aye-aye" refers to the entire animal, a part (e.g., claws), property (e.g., fur texture), movement, and so on. That children in fact solve this problem readily suggests that children must have some relevant capacities—be they dedicated word-learning constraints, social skills and theory of mind, domain-general associative mechanisms, or some combination of the above (Bloom, 2000; Golinkoff et al., 2000). Thus, the word-learning capacity of young children is a significant phenomenon that demands explanation.

Consider now the case of generic nouns. They include the same inductive puzzle as we see with specific reference, in that the referent is ambiguous (e.g., 'I like *aye-ayes*,' even in the context of an aye-aye, does not specify what aye-ayes are). Moreover, generics have additional properties that complicate the inductive learning problem children face. One most obvious complication is that the referent of a generic NP is not ever fully available in the naming context. Unlike the case of specific reference, where the referent is at least typically present (e.g., an aye-aye, which the adult speaker may point to), with generics the referent must be inferred (e.g., the class of aye-ayes, including past, present, future, and hypothetical exemplars).

The abstract nature of generics is particularly intriguing when one considers the standard description of children's early speech as focused on concrete referents, especially those that children can readily interact with (Nelson, 1973). Children's earliest words seem to reflect this principle (e.g., kitty, nose, shoe, banana, cookie, and juice are among children's earliest words; Fenson et al., 1994). I do not dispute this characterization; rather, the point is that this tendency renders acquisition of generics all the more impressive. Recently, some scholars have proposed that children are particularly likely to focus on individuals rather than categories (Sloutsky and Fisher, 2004), a tendency that, if true, should raise difficulties in acquiring generics.

Another complication is that although generics refer to kinds, they allow for exceptions (Carlson and Pelletier, 1995). We can say 'Dogs are four-legged' even though some dogs have only three legs. This point receives considerable attention in the linguistic and philosophical literatures, in part because it points to the nonobvious semantics of generics (e.g., Cohen, 2004; Leslie, 2007, 2008; Prasada, 2000). Some scholars argue, partly on the basis of this observation, that generics most likely do not simply map onto quantifiers (Carlson, 1999). A familiar example is 'Birds lay eggs,' which is a perfectly fine generic, whereas 'Birds are female' is not—even though most birds do not lay eggs (males, immature females) and all egg-laying birds are female. From the perspective of acquisition, the permissibility of exceptions could be doubly hard. On the one hand, it is an additional factor that must be mastered by children in order for them to use generics productively (how are they to know that it is okay to say 'Birds lay eggs,' but not okay to say 'Birds are female'?). On the other hand, it affects their capacity to solve the word-referent mapping problem. Thus, a child may hear 'Gold is yellow' but also see a brown lump that is labeled "gold" and have to reconcile apparently conflicting information.

The final complication I mention is the complexity of the mapping between form and meaning, in the expression of generics. (Here I focus exclusively on English, although the argument seems to hold for all languages [Carlson and Pelletier, 1995].) Simply put, there is no one-to-one mapping between form and meaning—far from it. Generics can be expressed with a variety of linguistic forms, and every kind of NP that is used to express generics can also be used to express nongenerics. For example, consider the following sentences:

(1) *Cats* are excellent mousers.
(2) *An elephant* has a long memory.
(3) *The artichoke* is chock-full of vitamins.

As can be seen in examples (1), (2), and (3), generic NPs can be expressed with a variety of forms (including bare plural ["birds"], indefinite singular ["an elephant"], and definite singular ["the artichoke"]).[1] Conversely, each of these NP forms can be used to express a nongeneric utterance, as shown in examples (4), (5), and (6):

(4) *Cats* were hanging around my garden last summer.
(5) *An elephant* tried to escape from the Toledo Zoo.
(6) *The artichoke* is on the kitchen counter.

---

1. Generics are not limited to these forms, as they can also be expressed with mass nouns ('Gold is valuable'), definite article plus adjective ('The rich get richer'), bare singular ('Peace on earth, good will to man'), and even definite plurals ('The Sioux Indians are a proud people').

This is not to say that formal cues are irrelevant, but rather that both formal (morphosyntactic) and contextual cues are needed to reach an interpretation of a sentence as generic or nongeneric.

To see the importance of multiple cues, consider the following sentences, with likely interpretation of "pizza" in each case marked in parentheses:

(7) Do you like *the pizza?* (nongeneric)
(8) Do you like *pizza?* (generic)
(9) Do you want *pizza?* (nongeneric)
(10) Would you like *pizza?* (nongeneric)
(11) Would you like *pizza*, if you were a dog? (generic)

We see that the only difference between (7) and (8) is the presence/absence of the word "the," and that the likely interpretation of the sentence varies accordingly: (7) refers to a particular pizza, whereas (8) refers to pizza in general. On the other hand, (8), (9), (10), and (11) all include the unmodified mass noun "pizza," yet the interpretation of this NP bounces back and forth between generic and nongeneric, depending on the verb ["like" (8) vs. "want" (9)], choice of auxiliary verb ["do" (8) vs. "would" (10)], or even the presence or absence of an additional phrase beyond the core sentence [(10) vs. (11)].

The lack of one-to-one mapping is hardly surprising, given that natural languages tend to be messy, redundant, and filled with exceptions—as a rule. Nonetheless, the exceptions are particularly striking in the case of generics. And from a developmental perspective, the complexity of cues certainly provides an additional challenge. Slobin (1973) has demonstrated that children generally learn regular systems more easily than systems that flout the principle of correspondence between form and meaning.

To summarize, then, generic NPs pose a potentially daunting task for child learners. In order to acquire generic NPs, children must hold in mind abstract referents that are not fully present in the naming context, recognize that generics are simultaneously broad in scope yet admitting of exceptions, and deal with formal linguistic cues that are varied and inconsistent.

Young Children's Early Use and Understanding of Generics

Despite the various challenges discussed above, children use and understand generics from a young age. In this section I review some of the evidence for this claim, as well as the implications for a model of learning. The evidence for children's early capacity to use generics comes from both production and comprehension studies, including children with varying types of language experience.

One rather straightforward approach to studying generic acquisition in children is to listen to children's spontaneous speech in natural settings. This approach has the advantage of being entirely neutral in context (i.e., not biasing toward generic production) and of enabling one to study the interplay between children's language and parents' language. Furthermore, researchers have suggested that natural conversations in a home setting are likely to be an especially sensitive index of children's language and conceptual abilities (Bartsch and Wellman, 1995). We therefore conducted a study of the developmental emergence of generics by examining longitudinal transcripts of parent–child conversations (Gelman et al., 2008). The data were provided by the CHILDES (Child Language Data Exchange System) project, organized by Brian MacWhinney and Catherine Snow (MacWhinney and Snow, 1990). The CHILDES database includes natural conversations that were recorded and transcribed by various child-language researchers, dating back more than thirty years. For our purposes, we included only English-speaking children between two and four years of age who contributed data to at least two of our three age groups: two years, three years, four years. Altogether we included eight children: four children at all three ages (Abe, Adam, Naomi, and Ross), two at ages two and three years (Nathaniel, Peter), and two at ages three and four years (Mark, Sarah). The researchers contributing these data were Kuczaj (1976), Brown (1973), Sachs (1983), MacWhinney (1991), Snow (see MacWhinney and Snow, 1990), and Bloom (1970).

We focused exclusively on transcripts for which children's mean length of utterance (MLU) was 2.5 and above for at least three taping sessions in a row, to ensure that children had command of the syntactic forms needed to express generics. For example, a child who does not yet productively distinguish singular from plural, or who does not yet reliably produce articles, cannot clearly indicate in English that a generic is intended. 'I like dog' could mean either "I like the dog" or "I like dogs." To reduce the coding task to more manageable size, we initially screened the data, limiting our analyses to utterances including plural NPs (e.g., 'dogs'), mass nouns (e.g., 'water'), or nouns preceded by "a" or "an" (e.g., 'a boy'). Based on pilot data as well as prior research (e.g., Gelman et al., 1998), we determined that generics in child-directed speech are mostly likely to appear in one of the above forms (e.g., definite NPs are rarely used to express generics in spontaneous child-directed speech). Each of these NPs was coded, in context, as generic or nongeneric. Using this method, we found that more than 8,000 generics were produced by the child and adult speakers.

The data are very rich and permit addressing a variety of questions concerning developmental and individual differences in children's production of generics. For current purposes, however, we touch on just two key findings. First, generics are frequent in children's natural speech. All six children for whom we had data at two years of age produced generics, as illustrated in the sentences below:

(12) "It's for *firemans*." (Abe)
(13) "Why he like '*paghetti*?" (Adam)
(14) "*Doggies* do poop." (Naomi)

(15) "Ride *trolleys*." (Nathaniel)
(16) "A blanket for *baby people*." (Peter)
(17) "*Skates* don't make a lot of noise." (Ross)

Perhaps most surprising, by four years of age children produced generics as frequently as did adults. The actual rate of generic production is roughly 2% of children's speech (increasing from about 1% at age two to about 3% at age four). Although this does not sound like a high rate, it must be considered in the context of all the other sorts of reference that children (and adults) are capable of producing. Nouns can function not only to make generic reference ('dogs are four-legged') but also to make singular definite reference ('that dog is four-legged'), general definite reference ('those dogs are four-legged'), distributive general reference ('every dog is four-legged'), collective general reference ('all dogs are four-legged'), specific indefinite reference ('I saw a four-legged dog'), and nonspecific indefinite reference ('I want to see a four-legged dog') (Lyons, 1977: 177–197). Given all these various functions, any given NP type will constitute only a small fraction of speech. Accordingly, even the most salient of NP types will occur in less than the majority of utterances. (Analogously, although food is a highly salient and important concept for young children, mention of food appears in much less than half of their utterances because there are many competing topics of conversation.) The rate of generics in our sample is comparable to the rate at which children produce genuine psychological references to thoughts and beliefs at six years of age (Bartsch and Wellman, 1995). Two percent of utterances still works out to several generics, on average, during the course of an hour of speech.

The second major point from these data is that children are not passively following in the footsteps of their parents, in their generic talk. Instead, they actively initiate generic conversations. One indirect piece of evidence for this point is that individual differences in parents' rate of generics do not reliably predict individual differences in the rate at which children produce generics. Children are highly consistent over time: the ordering of subjects from most to fewest generics (as percentage of utterances) is nearly identical at the three ages (e.g., Abe was the highest at every age; Peter was the lowest at every age). In contrast, adults show little to no consistency over time (e.g., Ross's parents showed one of the lowest rates when Ross was age two and one of the highest rates when he was age four). This would suggest that children's generics are not simply a function of the adult input.

Furthermore, we found that children at all three ages initiated generic talk. In order to address this issue, we examined all conversational sequences including generics. A conversational sequence was defined as a set of utterances focused on a given topic (e.g., bad people, houses, cookies). A conversational sequence could be as short as a single utterance, or more than 100 utterances in length. Our interest was in who initiated generic talk in these sequences: child or adult. For example, consider a hypothetical sequence in which the conversation at first focused on a particular cookie but then switched to talking about cookies in general. In this example, the person who switched the conversation to cookies in general was the one who initiated generic talk. Child-initiated sequences were those in which a child produced the first generic in the sequence; adult-initiated sequences were those in which an adult produced the first generic in the sequence.

Here is a sample adult-initiated sequence (with all target NPs in italics, and generics in bold italics), with Abe and his mother, for the category "elephant":

> ABE: It's *a elephant*.
> MOTHER: *A elephant?*
> ABE: Uh-huh.
> MOTHER: Do you like ***elephants***?
> ABE: Uh-huh. We seed *one* at the zoo.
> MOTHER: We sure did, we saw *him* eating, didn't we?
> ABE: Uh-huh. Hay, *he* ate hay!
> MOTHER: Uh-huh. ***Elephants*** like hay.
> ABE: Uh-huh, and peanuts, we getted some peanuts for *him*.
> MOTHER: That's right. Next time we go to the Chicago Zoo, maybe we'll see *him* again, *he*'ll probably remember you, because ***elephants*** never forget.

Here is a sample child-initiated sequence (for category "mommy"), with Abe and his father:

> ABE: Uh-huh that is for you, and we make a raccoon mask for me, and *Mommy* make a cow mask for *her*.
> FATHER: Is *Mommy* going to get a raccoon mask?
> ABE: No *her* will get a cow mask for *her*.
> FATHER: A cow mask?
> ABE: Uh-huh.
> FATHER: Does *Mommy* look like a cow?
> ABE: No, ***mommies*** don't look like cows, ***they*** have to get a mask.
> FATHER: *Mommy* needs a mask to look like a cow?

It is important to recognize that initiation rate is potentially independent of generic frequency. For example, it could be that all of young children's generics trail behind those of a parent. In other words, although children produced a substantial number of generics, the rate at which they *initiated* generics could be as low as 0% (or as high as 100%). However, what we found is that children initiated generics at a high rate, with four-year-olds initiating generics fully as often as adults. Even two-year-olds initiated generics roughly one-fourth of the time. Moreover, if we look just at sequences that include a child generic, most are child initiated.

This study of generics in natural parent–child conversations reveals an early capacity to produce generics in a typically developing sample of children. Obviously, for these children there is ample support for the generics they produce, in that the input language provides a rich source of information, examples, and ideas that children can exploit. This leads to the question of whether children are capable of producing generic language even in the absence of a richly supportive database–that is, when there is only minimal language input.

One way to examine this question is to study deaf children who are not provided with any experience with formal sign language (e.g., American Sign Language, or ASL). Because of their profound deafness and the lack of exposure to a gestural language, these children have little to no natural language input. Susan Goldin-Meadow and her colleagues have studied this population extensively, and found that such children typically invent their own symbolic gestural system, which Goldin-Meadow calls "home sign" (Goldin-Meadow, 2003). For example, a child might invent a sign to mean "grape" and a sign to mean "eat" and combine them into a two-gesture sequence meaning "(I) eat grapes." These home sign systems are revealing of what children are capable of expressing on their own, without explicit adult modeling, teaching, or feedback.

We examined videotapes of these children's spontaneous home signs, with both an American and a Chinese sample, and discovered a small but stable set of gestures that could be considered generic (Goldin-Meadow et al., 2005). We applied three criteria to code a gesture as generic. First, it had to reflect a general characteristic rather than an idiosyncratic one. For example, consider the hypothetical case of a deaf child reacting to a picture of a pig wearing a hat. If the child were to produce a *tail* gesture (a characteristic of all pigs), the gesture could be classified as a generic if it met the other two criteria. In contrast, if the child produced a *hat* gesture (a characteristic of that particular pig), the gesture could not be considered a generic under any conditions.

Second, the characteristic reflected in the gesture could not be present in the context. Thus, the *tail* gesture in the above example would not, in the end, be coded as generic if the tail was available in the picture—only if the tail was not present in the picture. The third criterion that a gesture had to meet to be classified as generic was that it had to refer not to a particular object but to a class of objects. In order to identify generics, we therefore relied heavily on contextual cues, both from the discourse and from our knowledge of the available objects and pictures (as in the tail example, above).

Note that we can never really be certain that a deaf child is expressing a generic. In the example above, the gesture could either mean "pigs have tails" or "this pig could/should have a tail." Whereas the former is a generic referring to pigs as a class, the latter is nongeneric, referring to a particular individual pig. In either case, however, the speaker is going beyond the specifics of the immediate situation and is making use of kind-based (generic) knowledge. That is, 'this pig could have a tail' expresses a property that is known only by reference to the generic category of which this individual is an instance.

Using this coding scheme, we found that six of the eight deaf children produced generics in their gestures, with no differences between the two cultures (see figure 6.1). For example, one child produced an "eat" gesture in response to a picture of a squirrel in a tree with nuts nearby, although no eating was taking place in the picture. We glossed this gesture as "squirrels eat nuts." Generic production rates were similar to those found in hearing adults (Gelman and Tardif, 1998). Furthermore, deaf children used their generics to refer to animals reliably more often than they used them to refer to artifacts, thus displaying the pattern found in prior research with hearing children and adults (Gelman, 2003; Gelman and Tardif, 1998; Gelman et al., 1998). This bias cannot be attributed to overall rate of talk about animals versus artifacts, because children did not display this animacy bias for nongeneric gestures. Thus, the animacy bias in generic usage does not need linguistic input to develop and is likely to reflect the different conceptual organizations that underlie these categories.

Taken together, these findings suggest that children readily and actively consider generic concepts and seek ways to express them symbolically. They do not require a model in order to produce generics.

To summarize the evidence from production data, generics are frequent in children's speech. Moreover, children are not simply following parental lead, because they initiate generics at a high rate. Generics also seem to be produced even with minimal language input (deaf children of hearing parents). Children therefore are spontaneous in their attention to generics.

These findings clearly demonstrate that children produce utterances that adults interpret as generic. But do children actually understand generics?

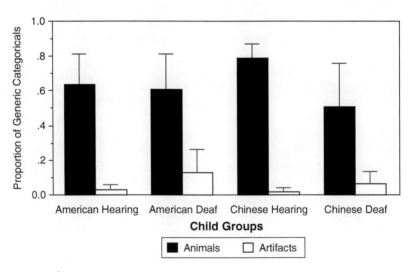

FIGURE 6.1 Generics in deaf and hearing children from two cultures (from Goldin-Meadow et al., 2005)

This is a nontrivial question, because it is certainly possible that children produce sentences that sound like generics but in fact merely mimic their outward form without having the same meaning. For this reason, it is critically important to examine children's *comprehension* of generic sentences. When we do so, we see that for the most part, generics seem to function for preschool children much as they do for adults: to make reference to a larger category that extends beyond the current conversational context.

Testing children's comprehension of generics is much trickier than it might at first appear. This point has both methodological and theoretical implications. Adults can rather directly supply interpretations of generic and nongeneric sentences, thereby revealing what interpretation is called to mind. For example, Twila Tardif and I employed a task with English- and Mandarin-speaking adults, asking whether each of a string of sentences referred to "most or any" members of a category, or to "one or a few" members of a category (Gelman and Tardif, 1998). Adults had no difficulty with the task, showing great consistency within each language—and even considerable consistency across languages—in making these judgments. The distinction we asked adults to make is only a crude approximation of the generic/specific distinction, in part because some generics do seem to refer to only a few members (e.g., 'Ticks carry the West Nile virus'; Leslie, 2007, 2008). Nonetheless, we can gauge semantic interpretations from adults because they can effectively provide translations of generic and nongeneric sentences. Preschool-age children, however, are notoriously poor at providing metalinguistic judgments (Gleitman et al., 1972).

What options, then, are available for studying children's semantic interpretations in the absence of metalinguistic judgments? The standard approach in the semantic development literature is to provide sample instances and ask children to assess whether they can be named by the word in question. For example, children might first learn a novel word (e.g., "dax") applied to a novel object (e.g., garlic press) and then be asked to select which of a variety of other objects is also a "dax" (e.g., objects varying in color, size, shape, texture). In this way, we can determine which properties children link to the word "dax." Hundreds, perhaps thousands of scientific papers have used this method, revealing many insights regarding children's acquisition of semantics.

The problem is, however, that this method does not work for studying generics. All of the studies using the approach mentioned above have examined specific reference ('Can you find another dax?'; 'Is this a dax?'). Generics cannot be studied in this way, for the simple reason that kind concepts cannot be presented in a concrete form. One cannot show dogs (generically), or even a picture of dogs (generically); one can only show a particular set of dogs. Accordingly, it is not possible to determine whether children understand that 'I like dogs' refers to dogs *as a kind*, by means of an object- or picture-selection task.

Our strategy for dealing with this dilemma has been to turn the problem around by exploiting the fact that generics refer to instances beyond those present in the available context. Thus, although one cannot ask children to point to a picture of "birds" as a kind, one can ask simple questions about "birds" as a kind. Lakshmi Raman and I developed a method that can be used with young preschool children (Gelman and Raman, 2003). We presented children with pictures of atypical category instances (e.g., two birds that cannot fly; two square balloons) and then asked questions either in generic form (e.g., 'Do birds fly?'; 'Are balloons round?') or in nongeneric form ('Do the birds fly?'; 'Are the balloons round?'). Because the instances in the pictures are atypical, these items allow one to differentiate generic from specific interpretations.

Using this method, we have found that children make use of form-class cues to identify generics, by two and a half years of age. As shown in figure 6.2, both children and adults interpreted nongeneric questions ('Do the birds fly?') as referring to the items in the present context (in this example, answering "no") and interpreted generic questions ('Do birds fly?') as referring to the items generically (in this example, answering "yes"). The subtle cue of presence or absence of "the" was sufficient to shift children's interpretation (see also Cimpian and Markman, 2008).

Children also used contextual cues to identify generics (Gelman and Raman, 2003). We provided children and adults with a picture of a single atypical instance (e.g., a short-necked giraffe) and asked one of two questions, either nongeneric or generic. The nongeneric question was, for example: 'Here is a giraffe. *Does it have a long neck or a short neck?*' Here we predicted that children would interpret "it" as referring to the specific giraffe in the picture and respond "short neck." The generic question was, for example, 'Here is a giraffe. *Do they have long necks or short necks?*' Here we predicted that children would interpret "they" as referring to the generic

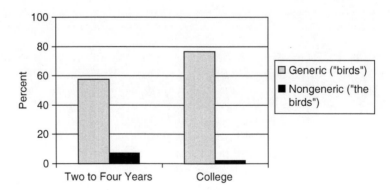

FIGURE 6.2 Mean percentage of categorywide responses, as a function of wording and age group (from Gelman and Raman, 2003)

kind of giraffes and respond "long neck." This strategy of making plural reference in the context of a single individual was one we noticed parents doing in their spontaneous speech with children (Gelman et al., 1998), for example:

"That's a chipmunk. And they eat the acorns."

As shown in figure 6.3, three-year-olds, four-year-olds, and adults all interpreted these two sentences differently, thereby showing implicit grasp of the generic–specific distinction as marked by contextual cues (i.e., the combination of wording and picture context). (Two-year-olds, however, did not perform differently from chance.)

We also conducted a control study, in which the identical question wording was used in both conditions ('Do they have long necks or short necks?'). However, we varied the picture, so that it either matched the plural wording (two short-necked giraffes) or mismatched the plural wording (one short-necked giraffe). Specifically, the generic condition was, "Here is a giraffe. Do they have long necks or short necks?" Because the pronoun "they" did not match the picture, we predicted that participants would interpret the word "they" as referring to giraffes as a kind, and respond "long necks." In contrast, the nongeneric condition was, "Here are two giraffes. Do they have long necks or short necks?" Because the pronoun "they" did match the picture, we predicted that participants would interpret the word "they" as referring to the giraffes in the picture and therefore respond "short necks." As shown in figure 6.4, this indeed is precisely what we found for three- and four-year-olds and adults.

Also using the approach of asking about predicates of generics and nongenerics, Michelle Hollander, Jon Star, and I found that English-speaking four-year-olds distinguish generics from both "all" and "some"

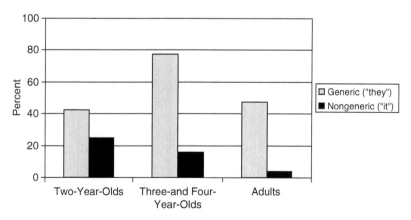

FIGURE 6.3 Contextual cues, mean percentage of categorywide responses, as a function of wording and age group (from Gelman and Raman, 2003)

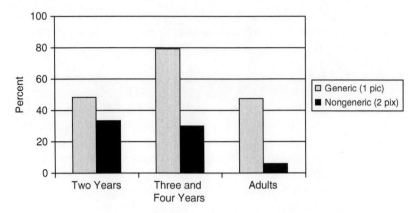

FIGURE 6.4 Contextual cues control, mean percentage of categorywide responses, as a function of wording and age group (from Gelman and Raman, 2003)

(Hollander et al., 2002). For example, children are much more likely to say "yes" in response to "Are some birds yellow?" than in response to "Are birds yellow?" They are also much more likely to say "yes" in response to "Are birds yellow?" than "Are all birds yellow?" It is remarkable that children honor these distinctions when one considers how closely generics overlap with both "all" and "some." On the one hand, generics could easily be confused with "all," as both typically express broad-scope, categorywide properties (e.g., '[all] dogs are furry,' '[all] bats fly around at night,' '[all] pencils are used for writing'). On the other hand, generics could be easily confused with "some," as both often employ bare plurals (e.g., 'I heard dogs barking last night,' 'bats live in that cave,' 'my teacher keeps pencils in her desk'). Despite these potential confusions, preschool children have untangled these different uses.

Summary

To summarize, children spontaneously begin to express generics between two and three years of age, and use generics as frequently as adults by four years of age. Even deaf children with minimal linguistic input fashion their own gestures to express generic concepts. This capacity to *produce* generics is matched by children's capacity to *comprehend* generics. Two-, three-, and four-year-olds are sensitive to subtle morphological cues (e.g., "the Xs" versus "Xs") to distinguish generic from specific reference. Additionally, three- and four-year-olds (but not two-year-olds) appropriately read contextual cues to determine that an utterance may be generic (e.g., interpreting a plural NP in the context of a single instance as signaling a generic intent). By four years of age, children distinguish generics from the quantifiers "all" and "some." These converging lines of evidence all suggest that preschool children have made remarkable headway on the problem of generic acquisition.

We do not mean to suggest that preschool children's generic use is indistinguishable from that of adults. To date, no evidence exists that young children use or understand less-common forms of generics (e.g., the definite singular: 'The dog is a noble beast'). Furthermore, in certain subtle respects, four-year-olds seem to interpret generics differently from adults (Gelman and Bloom, 2007). For example, adults interpret generics differently when referring to inborn properties versus acquired properties. Inborn properties (e.g., stripes that an animal is *born* with) are more likely to be predicated of a kind, even when all available instances have lost the property. In contrast, preschool children do not distinguish inborn from acquired properties in this respect. Nonetheless, the basic semantic implications of generics are in place by three to four years of age.

Models for Generic Acquisition

How can all this be accomplished? How do children learn the meanings of generic NPs? There is an ongoing debate in the literature concerning the kinds of abilities or mechanisms required for children to learn word meanings. Some have argued for innate, language-specific capacities; others have emphasized the role of social skills and theory of mind; still others have proposed that simple, domain-general associative mechanisms are sufficient (Golinkoff et al., 2000). I favor explanatory accounts that acknowledge multiple levels of skills and abilities (as argued in compelling detail in Bloom, 2000). But regardless of one's theoretical assumptions, a crucial point is that these debates until now have largely focused on how children learn to make *specific* reference (e.g., 'This is a dax'). Yet as we have seen, much of children's word learning involves generic reference. Indeed, I suspect that generics may be particularly important when introducing a new word (as definitions). A consideration of generic reference therefore provides an important new perspective on the issue.

Although I do not have a solution to the problem of how children learn generic NPs, I am skeptical that low-level associative mechanisms could provide a satisfactory answer. In prior studies of children's associative language learning, examples included a perceptually apparent semantic distinction that is rather directly linked to simple formal linguistic cues. For example, the count/mass distinction can be visually depicted[2] and is tightly correlated to formal cues (count vs. mass nouns). Smith et al. (1996:

---

2. This is true in most but not all cases (e.g., "furniture" is a mass noun; "plants" is a count noun). I thank an anonymous reviewer for this example.

145–146) propose that concepts are built by means of "dumb attentional mechanisms," as follows:

> [C]hildren repeatedly experience specific linguistic contexts (e.g., "This is a ___" or "This is some ___") with attention to specific object properties and clusters of properties (e.g., shape or color plus texture). Thus, by this view, these linguistic contexts come to serve as cues that automatically control attention.... [D]umb forces on selective attention—that is, associative connections and direct stimulus pulls—underlie the seeming smartness of children's novel word interpretations.

The problem is that generic NPs have no concrete perceptual manifestation and have no straightforward linguistic cues. By the time that children start to use generics productively (between two and three years of age), factors other than simple associative mechanisms must be guiding their word learning.

But how are generics acquired? At this time, the most plausible account seems to be that generics are a default interpretation for young children. First, there is no coherent set of generic markers in English, or in any other language, as far as we know (Carlson and Pelletier, 1995). As discussed above, formal cues to genericity are varied and inconsistent. Rather than there being a set of formal linguistic cues that signal that an utterance is generic, there seem instead to be nearly infinitely many cues (formal, pragmatic, semantic) that signal that an utterance is specific. For example, I can signal that an NP is specific by prefacing it with a possessive ('my dog'), pointing to a specific instance (e.g., pointing to a particular dog), using a deictic ('this dog'), situating the instance in a particular time point ('the dog woke me up in the middle of the night'), and so on. Generics may simply be anything that is left over—anything that is *not* marked as specific, in other words. Consistent with this position, generics in a number of languages (e.g., Mandarin or Quechua) are literally the unmarked form. For example, the following sentence in Mandarin

xiao3	ya1zi	yao2yao2bai3bai3	de	zou3	lu4
little	duck	waddlingly	DE	walk	road

could be translated in any of three different ways:

- The duck is waddling
- The ducks are waddling
- Ducks waddle. / A duck waddles.

Likewise, in Quechua, 'Huk'ucha chupayuq' could mean "The mouse has a tail," "Mice have tails," or "The mice have tails" (example thanks to Bruce Mannheim).

Thus, if I am correct, the task for children is not to acquire a list of all the ways that generics can be marked, but rather to learn to recognize when an utterance is specific. If children *assume* a conceptual distinction between generic and specific reference, then they can identify as generic those utterances that are not somehow marked specifically. It is in this sense that I propose generics as a *default*.

Implications of Generics for Other Cognitive Tasks

To this point we have considered generics as a window onto children's concepts, in that they are acquired readily despite challenging learning conditions. There is a further sense in which generics provide a window onto children's concepts as well. Namely, hearing generics in the input language may have implications for children's performance on other cognitive tasks, including word learning, inductive inferences, memory, and other forms of reasoning. This question is still in its infancy, so much of this section is speculative. My hope is to provide a context for considering these issues further.

I review the available evidence in terms of the kinds of effects that might be possible. I list several here, in order (arguably) from weakest to strongest:

- *Informing conceptual content.* Generics may help fill in the details of children's categories, either familiar or novel. Clearly, generics express propositions about a category (e.g., the statement 'Lions eat meat' expresses a proposition that a child may not have already known). Furthermore, hearing such properties in generic form may lead to a stronger link between the property and the category than if the property had been stated specifically (e.g., 'These lions eat meat'). Although this level of effect may seem fairly obvious, it is nontrivial, because it would require that generics lead to more powerful representations than nongenerics, that children understand generic input, that children retain generic form in long-term memory, and that they call this information to mind when new instances of the category are introduced.

    These sorts of conceptual content effects have been found in studies of preschool children's inductive reasoning. When four-year-old children learned novel properties about familiar categories, hearing the properties in generic form (e.g., 'Bears like to eat ants') yielded different inference patterns compared to when they heard the properties with either "some" (e.g., 'Some bears like to eat ants') or "all" (e.g., 'All bears like to eat ants') (Hollander et al., 2002). "All" elicited the most inferences, "some" elicited the fewest inferences, and generics were between "all" and "some." However, children made fewer category-based inferences from generics than did adults.

    Similar results are found when four-year-olds are asked to reason about novel categories. Hearing a novel property about a novel category in generic form (e.g., 'Pagons are friendly') led children to attribute the property to novel instances of the category more often than hearing the novel property in specific form (e.g., 'These pagons are friendly') (Chambers et al., 2006). Likewise, children are more likely to extend the novel word to instances that possess the property expressed in generic form (Hollander, Gelman, and Raman, 2009). For example, after seeing a single example of a "kevta" and hearing "Kevtas are woolly," four- and 5-year-old children are more likely to select a woolly but dissimilar animal as being a kevta than after hearing "This kevta is woolly." In contrast, children who hear "This kevta is woolly" are more likely to select a nonwoolly animal that is more similar overall to the initial instance.

• *Guiding attention.* Generics may guide children's attention toward different aspects of an object, category, or event. For example, Lakshmi Raman, Dedre Gentner, and I (Gelman et al., in press) are finding that children and adults make different sorts of comparisons, depending on whether the question they hear is posed generically ('Can you tell me some things that are different about *dogs and cats?*') or specifically ('Can you tell me some things that are different about *this dog and this cat?*'). Likewise, Michelle Hollander and I have some preliminary data with adults suggesting that framing a question in generic form ('Why do crabs hide under rocks?') encourages different sorts of explanations than framing a question in specific form ('Why is that crab hiding under a rock?'; Hollander and Gelman, 2006). Adults clearly have access to multiple ways of thinking about comparison and explanation, but the form of the question guides them to focus on different responses. More research is needed to determine whether children are sensitive to the same distinction in their explanations.

• *Identifying which categories are kinds.* Generics may further help children sort out which categories are relatively impermanent or arbitrary and which categories are more stable and inference-rich. When children first learn about a novel category, they may be uncertain what sort of grouping it is. Generics may provide an important cue. For example, suppose a child in the 1950s didn't know anything about politics but heard many generics from adults about "Communists." This input may have led to a certain representation of Communists (even in the absence of any particular facts) and to the assumption that members of this group are deeply alike in nonobvious ways.

This prediction falls from prior semantic analyses suggesting that generics express properties that are relatively more central to a category than properties that cannot be expressed with generics (Lyons, 1977). This notion is elaborated by Prasada and Dillingham (2006) in terms of "principled connections." Principled connections between a category and a property imply that we have a normative expectation that category members *should* possess the property in question and that category membership can be used to explain the presence of the property ('Tweetie lays eggs *because* Tweetie is a bird'). In a series of experiments, Prasada and Dillingham (2006) elegantly demonstrated the existence of principled connections in adults' commonsense reasoning about generic sentences. Nonetheless, Prasada and Dillingham also found that not all generics express principled connections. For example, 'Barns are red' is a perfectly good generic, because it expresses a true generalization about the category of barns, but it does not express principled connections. Instead, it conveys a statistical regularity (most barns are red).

• *Fostering a categorical perspective.* A potentially broader effect of generics may be to encourage a certain outlook about categories. Generics may encourage people to focus on a category-level of construal more than an individual-level of construal. Individual people seem to differ in the extent to which they stereotype, essentialize, or focus on an individual's category membership (Greenwald et al., 1998). It may be that hearing plentiful generics encourages this perspective. For example, a child who hears a great deal of generic talk in his or her input may be more readily disposed to making judgments more on the basis of category membership than individual identity. (Conversely, it may be that people who are more disposed to see the world in terms of categories use more generics—such that generics are an implicit index of conceptual outlook.) To date, there is some evidence for large variation in the amount of generic talk that both children and parents produce (Gelman, 2003) but no studies examining whether such variation is linked to stereotyping or essentializing.

• *Establishing kinds ontogenetically.* It is at least theoretically possible that generic language could permit children to establish kinds in the first place. However, I list this "potential" cause to illustrate what I see as the limits of language effects. I would argue that generics do not create an understanding of kinds (for a more extensive argument, see Gelman, 2003). As we have seen with the work with Goldin-Meadow et al. (2005), children with minimal language input nonetheless seem to work with kind concepts, and even to express them using their own invented communication system. Moreover, there is now good evidence that prelinguistic infants have some capacity to represent kinds (Waxman and Markow, 1995; Xu, 1999).

Conclusions

Despite the potential challenges that generics pose for acquisition, children learn and use generics from an early age. Moreover, children with minimal language input can express generic concepts. We therefore suggest that generics do not create kinds. Rather, children seem to possess kind concepts beforehand that permit early acquisition of generic NPs. Indeed, in one sense generics seem to be a default interpretation for young children. This result runs counter to the popular image of young children as limited to concrete concepts and incapable of abstract thought. It also calls into question the claim that "young children's naming of objects is principally a matter of mapping words to selected perceptual properties" (Smith et al., 1996: 144).

From a developmental perspective, generics are of interest not only as a problem space for young learners but also as a form of language that seems to have measurable effects on children's reasoning. At the very least, generics highlight certain properties for children. I have also sketched out some respects in which generics may have broader effects on cognition. At this point, research is only beginning to explore these issues, and many questions remain concerning the strength and scope of these effects, as well as how generic language interacts with preexisting knowledge and beliefs. One major unanswered question concerns whether generics merely focus children's attention in particular ways, or whether generics are capable of leading children to construct different kinds of concepts than they might otherwise do. In any case, the available evidence does suggest that this is a fruitful line of research.

ACKNOWLEDGMENT

This research was supported by National Institute of Child Health and Human Development grant HD-36043 and NSF grant BCS-0817128 to Susan A. Gelman.

REFERENCES

Bartsch, K., and Wellman, H.M. (1995). *Children talk about the mind.* Oxford: Oxford University Press.
Bloom, L. (1970). *Language development: form and function in emerging grammars.* Cambridge, MA: MIT Press.
Bloom, P. (2000). *How children learn the meanings of words.* Cambridge, MA: MIT Press.
Brown, R.W. (1973). *A first language: The early stages.* Cambridge, MA: Harvard University Press.
Carlson, G. (1999). Evaluating generics. *Illinois Studies in the Linguistic Sciences, 29,* 13–24.
Carlson, G.N., and Pelletier, F.J. (Eds.) (1995). *The generic book.* Chicago: University of Chicago Press.
Chambers, C.G., Graham, S.A., and Turner, J.N. (2006). When hearsay trumps evidence: How generics guide preschoolers' inferences about novel kinds. Unpublished manuscript, University of Toronto.
Cimpian, A., and Markman, E.M. (2008). Preschool children's use of cues to generic meaning. *Cognition, 107,* 19–53.
Cohen, A. (2004). Generics and mental representations. *Linguistics and Philosophy, 27,* 529–556.
Fenson, L., Dale, P.S., Reznick, J.S., Bates, E., Thal, D.J., and Pethick, S.J. (1994). Variability in early communicative development. 13–24. *Monographs of the Society for Research in Child Development,* Serial 242, Vol. 59, No. 5.

Gelman, S.A. (2003). *The essential child: Origins of essentialism in everyday thought*. New York: Oxford University Press.

Gelman, S.A., and Bloom, P. (2007). Developmental changes in the understanding of generics. *Cognition, 105*, 166–183.

Gelman, S.A., and Raman, L. (2003). Preschool children use linguistic form class and pragmatic cues to interpret generics. *Child Development, 74*, 308–325.

Gelman, S.A., and Tardif, T.Z. (1998). Generic noun phrases in English and Mandarin: An examination of child-directed speech. *Cognition, 66*, 215–248.

Gelman, S.A., Coley, J.D., Rosengren, K., Hartman, E., and Pappas, T. (1998). Beyond labeling: The role of parental input in the acquisition of rightly-structured categories. *Monographs of the Society for Research in Child Development*, Serial 253, Vol. 63, No. 1.

Gelman, S.A., Goetz, P.J., Sarnecka, B., and Flukes, J. (2008). Generic language in parent-child conversations. *Language Learning and Development, 4*, 1–31.

Gelman, S.A., Raman, L., and Gentner, D. (in press). Effects of language and similarity on comparison processing. *Language Learning and Development*.

Gleitman, L.R., Gleitman, H., and Shipley, E.F. (1972). The emergence of the child as grammarian. *Cognition, 1*, 137–164.

Goldin-Meadow, S. (2003). *The resilience of language: What gesture creation in deaf children can tell us about how all children learn language*. New York: Psychology Press.

Goldin-Meadow, S., Gelman, S.A., and Mylander, C. (2005). Expressing generic concepts with and without a language model. *Cognition, 96*, 109–126.

Golinkoff, R.M., Hirsh-Pasek, K., Bloom, L., Smith, L.B., Woodward, A.L., Akhtar, N., Tomasello, M., and Hollich, G. (2000). *Becoming a word learner: A debate on lexical acquisition*. New York: Oxford University Press.

Greenwald, A.G., McGhee, D.E., and Schwartz, J.L. K. (1998). Measuring individual differences in implicit cognition: The implicit association test. *Journal of Personality and Social Psychology, 74*, 1464–1480.

Hollander, M., and Gelman, S.A. (2006). Causal explanations in the context of generic and non-generic language. Unpublished data, University of Michigan.

Hollander, M., Gelman, S.A., and Raman, L. (2009). Generic language and judgments about category membership: Can generics highlight properties as central? *Language and Cognitive Processes, 24*, 481–505.

Hollander, M.A., Gelman, S.A., and Star, J. (2002). Children's interpretation of generic noun phrases. *Developmental Psychology, 38*, 883–894.

Kuczaj, S. (1976). -ing, -s and -ed: A study of the acquisition of certain verb inflections. Unpublished doctoral dissertation, University of Minnesota.

Leslie, S.-J. (2007). Generics and the structure of the mind. *Philosophical Perspectives, 21*, 375–405.

Leslie, S.-J. (2008). Generics: Cognition and acquisition. *Philosophical Review, 117*, 1–49.

Lyons, J. (1977). *Semantics*, Vol. 1. New York: Cambridge University Press.

MacWhinney, B. (1991). *The CHILDES project: Tools for analyzing talk*. Hillsdale, NJ: Erlbaum.

MacWhinney, B., and Snow, C. (1990). The child language data exchange system: An update. *Journal of Child Language, 17*, 457–472.

Markman, E.M. (1989). *Categorization and naming in children: Problems in induction*. Cambridge, MA: Bradford Book/MIT Press.

Nelson, K. (1973). Structure and strategy in learning to talk. *Monographs of the Society for Research in Child Development*, Serial 149, Vol. 38, Nos. 1 and 2.

Prasada, S. (2000). Acquiring generic knowledge. *Trends in Cognitive Sciences*, 4, 66–72.

Prasada, S., and Dillingham, E.M. (2006). Principled and statistical connections in common sense conception. *Cognition*, 99, 73–112.

Quine, W.V. (1960). *Word and object*. Cambridge, MA: MIT Press.

Sachs, J. (1983). Talking about the there and then: The emergence of displaced reference in parent-child discourse. In K.E. Nelson (Ed.), *Children's language* (Vol. 4, pp. 1–28). Hillsdale, NJ: Erlbaum.

Slobin, D.I. (1973). Cognitive prerequisites for the development of grammar. In C.A. Ferguson and D.I. Slobin (Eds.), *Studies of child language development* (pp. 175–208). New York: Holt, Rinehart and Winston.

Sloutsky, V.M., and Fisher, A.V. (2004). When development and learning decrease memory: Evidence against category-based induction in children. *Psychological Science*, 15, 553–558.

Smith, L.B., Jones, S.S., and Landau, B. (1996). Naming in young children: A dumb attentional mechanism? *Cognition*, 60, 143–171.

Waxman, S.R., and Markow, D.B. (1995). Words as invitations to form categories: Evidence from 12- to 13-month-old infants. *Cognitive Psychology*, 29, 257–302.

Xu, F. (1999). Object individuation and object identity in infancy: The role of spatiotemporal information, object property information, and language. *Acta Psychologica*, 102, 113–136.

PART II

MASS TERMS

7

Mass Terms: A Philosophical Introduction

FRANCIS JEFFRY PELLETIER

On an intuitive level, mass nouns are those such as[1]

(1) *water, cutlery, lamb* (the food), *spaghetti, mud, beer, gold, equipment, software, hardware, cheese, oats,* ....

As well, many "abstract" nouns are mass, for example,

(2) *trust, help, intelligence, information, damage, knowledge,* ....

These are contrasted with "concrete" count nouns such as

(3) *dog, tree, father, piano, lamb* (the animal), *biscuit, noodle, prize, child, knee,* ...,

as well as to some "abstract" count nouns, for example,

(4) *failure, belief, proposal, problem,* ....

Complex phrases are also brought into the categorization, so that 'cutlery that is in the drawer' and 'hot water' are mass phrases while 'tree that is in the park' and 'tall person' are count phrases. I use "mass term" and "count term" to cover both nouns and more complex noun phrases.

---

1. The issues discussed in this overview are developed in more detail in Pelletier and Schubert (2003).

(As well, some theorists admit other grammatical categories into the count/mass realm: some verb phrases and some adjectives have been argued to be mass/count, but I won't follow up that line of thought in this overview.)

When one focuses on the syntactic aspects of mass versus count, we are told that count terms syntactically admit numeral modifiers and quantifiers that presuppose a method of counting but mass terms do not, and related to this is the fact that count terms but not mass terms can be pluralized:

(5) *three dogs, a thousand trees, several failures,* ...
(6) **three waters, *a thousand cutleries, *several helps,* ....

Mass terms are thus always singular[2] and have their own method of measurement that is not appropriate with singular count terms:

(7) *much water, a lot of cutlery, little knowledge,* ...
(8) **much dog, *a lot of tree, *a little belief.*[3] ...

From a semantic point of view, the fundamental difference between mass and count terms is that count terms are true of *objects*—entities that are distinct from each other and thus one can distinguish and count them— while mass terms are true of *stuff* that is undifferentiated with respect to the term being used to describe it. Mass terms are therefore unlike count terms in that they are *divisive* in their reference: they permit something that the mass term is true of to be arbitrarily subdivided and the term to be true of these parts as well. Taking the water in the glass to be something that 'is water' is true of, it can be arbitrarily divided into parts and 'is water' will be true of both parts. And again, mass terms, unlike count terms, are also *cumulative* in their reference: putting the water contained in two glasses into a bowl yields something of which 'is water' is true. But the same is not the case with a count term like *dog*. Chopping up a dog into parts arbitrarily does not yield more things of which 'is a dog' is true, nor do two dogs make a thing of which 'is a dog' is true.[4]

The foregoing was a rough-and-ready, intuitive description of what mass terms are. This is perhaps a sufficient background for understanding many of the points made in the chapters of this part of the volume, but there are many further issues that remain outstanding in linguistico-philosophical theory of mass terms. And so I would like to take this opportunity to delve deeper.

---

2. With some exceptions, such as *oats* and *smarts*.
3. *Little* in (7) and (8) is the measure term, not a size- (or importance-) indicating adjective.
4. Other than in a Frankenstein-like scenario.

The Traditional Account

First, I expand on the examples just given to show how the background theory works. After this I consider a group of objections that seem to show that the entire theory is wrong. And it is here where linguistics and philosophical semantics would like some "outside assistance."

As with the earlier generics overview (Chapter 1), I use terminology of "noun" (N), "common noun" (CN), and "noun phrase" (NP), respectively, to designate (certain) lexical items, some adjectival and relative clausal modifications of nouns, and a CN that has been "determined" by the addition of a (definite or indefinite) article, or a demonstrative, or a quantifier. Also relevant especially to discussions of mass terms (and also plurals) are what are known as "classifier phrases." In English these usually take the form "X of," which is then applied to a mass term. I consider the result to be a type of CN. For example,

(9) a. bowl of jelly
     b. blade of grass
     c. puddle of water

and the like. In turn these can be converted to NPs by the addition of determiners and quantifiers, so that we have[5]

(10) a. the bowl of jello
     b. this blade of grass
     c. a puddle of water

A standard way to view nouns (e.g., Quirk et al. 1985; Huddleston and Pullum 2002) is to say that they come with certain syntactic features plus a semantic value. In describing larger and larger phrases that contain the noun (as it occurs in a sentence), the syntactic features are employed so as to guarantee that well-formedness conditions are met by these larger phrases, and their semantic features are generated in some suitable way. For example, *boy* might be syntactically characterized as an N that is singular and masculine, with a semantic value of the set of all individual boys;[6] *smart* might be syntactically characterized as an adjective with a semantic value of being a function that selects the smart objects out of a given set of objects. Then the complex phrase *smart boy* could be syntactically characterized as a CN that is singular and masculine, and its semantic value would be the set of all individual smart boys. If we now tried to add the determiner/quantifier *many* to this CN so as to form a full NP, we discover that it fails because *many* has a syntactic requirement that it requires a nonsingular CN as an argument. And hence **many smart boy* is

---

5. There is more to this syntactic story, but this is enough for the purposes of this overview. See Quirk et al. (1985, §§5.6–5.8) or Huddleston and Pullum (2002, §3.1).
6. The semantic value is only for the purposes of this example.

syntactically ill-formed (and the question of its semantic value doesn't even arise). Using *the* to form the full NP, however, *would* be syntactically appropriate, and the semantic value of *the smart boy* is the most salient smart boy in the relevant context.[7] If there is no such item, then the sentence in which this NP occurs is semantically anomalous or maybe false (depending on the theory), but it retains its syntactic good standing. In this general sort of view, the semantic value of complex terms (CNs and NPs) that contain mass or count nouns as parts are computed as some function of the semantic value of the embedded noun, the particular function depending on what the other parts of the complex are. Without involving ourselves in details of just exactly which functions are used for which syntactic combinations, we can give examples, such as: "The semantic value of *dirty water* is describable as, or computed in accordance with, whatever the semantic value of *water* is, and whatever the semantic value of *dirty* is, when they are put together by the syntactic rule of an adjective modifying a noun to form a CN." This general account of how the syntactic well-formedness constraints work with the semantic values of syntactically simple pieces of language to construct the semantic values of the syntactically more complex items is called "semantic compositionality" and is a touchstone for most modern semantic theories.

Applying this picture to mass terms goes as follows. Some nouns in the lexicon are marked +count while others are marked +mass.[8] In this picture, +mass is a syntactic feature that enforces certain well-formedness constraints so that the asterisked phrases in (6) and (8) are classified as violations of *syntactic* well-formedness constraints, and thus join

(11) *Dog the quickly

in being simply ungrammatical, and thus without any semantic value.

Problems for the Traditional Account

The traditional account is syntax driven. As the account given above showed, the lexical items are assigned either a +mass or +count feature, and this feature controls the syntactic admissibility or inadmissibility of larger phrases. But there are many words that have both mass and count meanings, for instance,

(12) a. Concrete terms
 (i) a lot of chocolate / many more chocolates
 (ii) play football / buy a football

---

7. Again, the semantic value is just for expository purposes.
8. In many places, for example, Quirk et al. (1985) and Huddleston and Pullum (2002), −count is used in place of +mass. But we will continue with the terminology that is more common in philosophy and psychology.

(iii) too much paper / write a paper
(iv) drink beer / drink a beer

b. Abstract terms
(i) much discussion / three different discussions
(ii) much justification / many justification
(iii) a lot of difference / many differences
(iv) much more data / many more data

The examples in (12) are just the tip of the iceberg. There are many more of the "dual life" terms that have been illustrated in (12), and they seem to form regular patterns:

(13) Mass terms used "countily":
a. Pinot Noir is *wine* / Pinot Noir is *a wine*
b. Kim produces *sculpture* / Kim is producing *a sculpture*
c. Sandy likes *lamb* / Sandy likes *every lamb*
d. *Beer* on the table / Three *beers* on the table / Eight *beers* on tap

(14) Count terms used "massily"
a. Leslie has more *car* than *garage*
b. Chris Pronger, 6'6" worth of Anaheim Duck *defenseman*...
c. He's got *woman* on his mind
d. What a hunk of *man*!
e. Some people like *data* better than *theory*

And then there's the "universal grinder" of Pelletier (1975), which is like a meat grinder except that it can accommodate any object, no matter how large, and its teeth are so powerful and fine that it can grind anything, no matter how strong. One inserts an object that falls under any (concrete) count noun into one side, for example, a hat. Push the button, and the result is that there is hat all over the floor.[9] Another push of the button and we can have book all over the floor. An unfortunate accident might generate curious cat all over the floor.

One might also think of "universal packagers" in this regard, which take any item of which a mass term is true and convert it into an object. Any time there is a use for a particular type or amount of some mass, then there can be a count term that describes it, for example, *a* 'finely silted mud,' which can be a name for a type of mud and also a predicate that is true of all individual exemplars of this type. And if there is a standardized amount of *M* that is employed in any use, then there will be a count term that describes this amount, such as 'a beer' or 'an ice cream.' In fact, there seems always to be a count use for any alleged mass term *M*, meaning (roughly) "a kind of *M*." Putting all these together, a term like "a scotch" could be true of

---

9. This is true despite the fact that we might have some other term, for example, *felt*, that also describes what is on the floor.

individual servings (thus being independently true of each piece of the actual matter in the various glasses), or true of the different amounts (so that two instances of the same one-ounce serving count as only one amount), or true of the different kinds of scotch on the table.

These considerations show that the appropriate theory needs to talk about *meanings* of terms, or *uses* of the terms, or maybe *occurrences* thereof (some occurrences are mass, others of the same word are count). But then this is no longer a syntactic account! And the traditional theory just doesn't work. It will turn out that since *any* noun can be either mass or count, a +mass/+count syntactic distinction does no work—*nothing* is ruled out by the syntactic rules.

So we need to find some alternative approach.

A Semantic Approach?

A semantic approach would be one that interpreted the features +mass and +count as descriptions of the semantic value of Ns and CNs and so on. Thus, they would not figure in the syntactic well-formedness constraints of a grammar, but would emerge as a description of what the semantic values of the embedded nouns are, and how these semantic values get altered by the syntactic combination of those nouns with other words. In such a picture, the features do not syntactically rule anything out; the most that can be said is that certain combinations are "semantically anomalous" and hence can't be interpreted.

As shown above, many words have both a natural mass and a natural count sense, so the basic thing that gets entered into a phrase structure description of a sentence will be one of these senses. It is never very clear how this is supposed to be effected in a grammar, but we will not pause over that here, and simply assume that there is some way that this can be done. But even if we can assume this, there nonetheless seem to be some serious difficulties that are mirrors of the syntactic approach given above.

We have already seen what theorists have asserted about the semantics of mass terms, in contrast to count terms:

(15) a. Mass terms are *true of stuff*
 b. Mass terms are *divisive in their reference*
 c. Mass terms are *cumulative in their reference*
 d. Stuff that mass terms are true of *cannot be counted*.
 e. Stuff that mass terms are true of *can be measured*

Some theorists take the divisiveness and the cumulativity conditions together to be called the "homogeneous in reference" condition. Many formal semanticists (e.g., Link 1983; Chierchia 1998a, 1998b; Pelletier and Schubert 2003) take the characteristics in (15) to be best accounted for in terms of a semilattice theory. A semilattice has no lowest elements and is atomless. The idea is that anything that *water*, for example, might be true of has subparts—things in the lattice that are its parts—of which *water* is true; and any two elements in the *water* lattice find a joined element also in the lattice that represents the merge of those two elements.

Problems for a Semantic Approach

Many mass terms obviously are not "atomless" in the sense required by this theory. Consider

> (16) *furniture, cutlery, clothing, equipment, jewelry, crockery, silverware, footware....*

Clearly there are atomic parts of these, and yet they are considered mass terms by any of the traditional grammars. So it cannot be an atomless mereology that accounts for the mass nature of these words, and by extension, since it doesn't account for the mass nature of these particular words, there seems to be no reason to think it accounts for the mass nature of *any* words.

In any case, perhaps we should further examine the presumption that there are any words at all that obey the condition on divisiveness. Or, put another way, are there really any words that are atomless—that have no smallest parts? Doesn't *water*, for example, have a smallest parts: H_2O molecules perhaps? Certainly *coffee* has smallest parts, as do other mixtures. A standard defense of the divisiveness condition in the face of these facts is to distinguish between "empirical facts" and "facts of language." It is an empirical fact that water has smallest parts, it is said, but English does not recognize this in its semantics: the word *water* presupposes infinite divisibility.

It is not clear that this is true, but if it is, the viewpoint suggests interesting questions about the notion of semantics. If *water* is divisive but water isn't, then water can't be the semantic value of *water* (can it?). In turn, this suggests a notion of semantics that is divorced from "the world," and so semantics would not be a theory of the relation between language and the world. But it also would seem not to be a relation between language and what a speaker's mental understanding is, since pretty much everyone nowadays *believes* that water has smallest parts. Thus, the mental construct that in some way corresponds to the word *water* can't be the meaning of *water* either. This illustrates a kind of tension within "natural language metaphysics." This very puzzling state of affairs is something that theoreticians from philosophical and linguistic semantics would welcome clarification about from psychological studies of the sort carried out in the present conference.

Further problems with any semantic approach to the mass–count distinction comes from the fact that such theories would be forced to admit that there are pairs of words where one is semantically mass and the other is semantically count and yet the items in the world that they describe seem to have no obvious difference that would account for this. On the intuitive level, it seems that postulating a *semantic* difference should have some reflection in the items of reality that the terms designate. But this is just not true. There seems to be nothing in the *referent* of the following mass versus count terms that would explain how they should be distinguished—as they intuitively are.

(17) a. Concrete terms
        (i) *baklava* versus *brownies*
        (ii) *spaghetti* versus *noodles*
        (iii) *garlic* versus *onions*
        (iv) *rice* versus *beans*
    b. Abstract terms
        (i) *praise* versus *compliments*
        (ii) *knowledge* versus *beliefs*
        (iii) *flu* versus *colds*
        (iv) *information* versus *facts*

And along the same lines, one and the same item of reality can be referred to by means of a count term in one language and a mass term in another:

(18) a. *dandruff* versus *les pellicules*
     b. *dishes* versus *la vaisselle*

The cross-linguistic facts are much more widespread and intricate than these few examples between French and English might suggest. Some discussion is in Bunt (1985), Krifka (1995, 1998), Chierchia (1998a) Borer (2005), and many other places. For instance, Chierchia (1998a) mentions that, even though Italian matches English in having both a mass noun corresponding to "hair" (*capello*) and a count noun corresponding to "hairs" (*capelli*), in English one says

(19) a. I cut my hair
     b. *I cut my hairs

while in Italian one says

(20) a. *Mi somo tagliato i capello
     b. Mi somo tagliato i capelli

It would seem that the same activity is described in the two cases, so there can't really be anything in the choice of mass versus count.

How can all these observations be reconciled? That is an important question that philosophical and linguistic semanticists would like to have some input from psychological studies, both of the adult behavioral type and of the developmental sort. Perhaps a wider net of collaborators with a wider range of phenomena under investigation can yield a pleasing answer. At least, that is a hope for the present volume and the research being reported.

REFERENCES

Borer, Hagit. *Structuring Sense*. Vol. 1: *In Name Only*. Oxford University Press, Oxford, 2005.

Bunt, Harry. *Mass Terms and Model Theoretic Semantics*. Cambridge University Press, Cambridge, 1985.

Chierchia, Gennaro. Reference to kinds across languages. *Natural Language Semantics*, 6: 339–405, 1998a.

Chierchia, Gennaro. Plurality of mass nouns and the notion of "semantic parameter." In S. Rothstein, editor, *Events and Grammar*, pages 53–103. Kluwer, Dordrecht, 1998b.

Huddleston, Rodney, and Geoffrey Pullum. *The Cambridge Grammar of the English Language*. Cambridge University Press, Cambridge, 2002.

Krifka, Manfred. Common nouns: A contrastive analysis of English and Chinese. In Gregory Carlson and Francis Jeffry Pelletier, editors, *The Generic Book*, pages 398–411. University of Chicago Press, Chicago, 1995.

Krifka, Manfred. The origins of telicity. In S. Rothstein, editor, *Events and Grammar*, pages 197–236. Kluwer, Dordrecht, 1998.

Link, Godehard. The logical analysis of plurals and mass terms: A lattice-theoretical approach. In R. Baüerle, C. Schwarze, and A. von Stechow, editors, *Meaning, Use and Interpretation*, pages 303–323. de Gruyter, Berlin, 1983.

Pelletier, Francis Jeffry. Non-singular reference: Some preliminaries. *Philosophia*, 5: 451–465, 1975. Reprinted in Francis Jeffry Pelletier, editor, *Mass Terms: Some Philosophical Problems*, pages 1–14. Kluwer Academic, Dordrecht, 1979.

Pelletier, Francis Jeffry, and Lenhart Schubert. Mass expressions. In F. Guenthner and D. Gabbay, editors, *Handbook of Philosophical Logic*, 2nd Ed., Vol. 10, pages 249–336. Kluwer, Dordrecht, 2003. Updated version of the 1989 version in the 1st edition of *Handbook of Philosophical Logic*.

Quirk, Randolph, Sidney Greenbaum, Geoffrey Leech, and Jan Svartvik. *A Comprehensive Grammar of the English Language*. Longman, New York, 1985.

8

A Piece of Cheese, a Grain of Sand: The Semantics of Mass Nouns and Unitizers

CLIFF GODDARD

1. Introduction

In her classic paper "Oats and Wheat: Mass Nouns, Iconicity and Human Categorization," Anna Wierzbicka (1988) argued the case for the existence of numerous, subtly different, subclasses of mass nouns and postulated detailed explanatory links between underlying conceptualizations and grammatical behaviors. She also stressed the partly language-specific character of these subclasses and suggested that differences between languages are often related to culture (e.g., connected with different eating and food preparation practices). In this study, I aim to extend and improve on Wierzbicka's arguments and analyses, concentrating on concrete mass nouns in English. The two overriding points of the entire study are that the formal linguistic properties of mass nouns are systematically correlated with their conceptual content, and that this conceptual content can be identified with rigor and precision using appropriate methods of linguistic semantics. The analytical framework is the Natural Semantic Metalanguage (NSM) system of lexical semantic representation (Goddard and Wierzbicka 2002; Wierzbicka 1996).

Section 2 proposes a formal semantic schema for the shared semantic components of prototypical count nouns. Section 3 addresses a sample of seven different subclasses of concrete mass nouns, typical members being, respectively, *cheese, wheat, dust, gravel, oats, coffee grounds,* and *noodles.* Sections 4 and 5 explicate the semantics of "unitizer" constructions of several kinds, for example, 'a piece of cheese,' 'a grain of sand.' Conclusions and reflections form section 6. (In the appendix I show that English-specific rules of regular polysemy are needed to account for the conversion of count nouns to mass nouns, in sentences like 'I don't eat tomato' and 'There was cat all over the driveway.')

There has been a lot of discussion about what, if anything, "mass expressions" all have in common (see Pelletier and Schubert 2003). The picture that emerges very clearly from Wierzbicka (1988) is that, even excluding abstract words such as *justice, time, heat, advice,* and *mathematics,* the term "mass noun" (or "mass expression") does not designate a category in the true sense but is best regarded as the converse of "count noun" (see Payne and Huddleston 2002: 340). This is as one might expect, given that most of the standard criterial properties for mass nouns are negative, such as the absence of the singular/plural contrast, incompatibility with numerals and similar quantifiers, and inability to occur with pronominal *one* as an anaphor.[1] There can be little semantic homogeneity across groupings as diverse as the following (and there are still others): (a) kinds of concrete physical things, such as *water, sand,* and *oats*; (b) functional-collective words such as *fruit, vegetables, furniture,* and *crockery*; (c) words such as *leftovers, belongings,* and *remains* that refer to things of various kinds united by spatial and temporal contiguity; (d) place-related nouns, such as *stairs, ruins, steppes, shallows,* and *woods*; (e) some internal body-part words, such as *guts, bowels,* and *brains,* with multiple undifferentiated parts; (f) certain illnesses and bodily conditions with a multiple aspect, such as *mumps, sniffles, measles, scabies, goosebumps, pins and needles*; and (g) "dual objects," such as *scissors, scales, panties,* and *headphones.* For reasons of space, the present study addresses only examples of the first of these types, which I will refer to as concrete mass nouns.

Even under the rubric of concrete mass nouns, a considerable number of different subclasses can be identified on the basis of variable linguistic properties, such as

- whether the noun occurs in the singular or in the plural form, for example, *wheat* versus *oats*;
- whether the referent has recognizable parts or units and, if so, whether these are named, for example, 'grains of rice,' 'flakes of snow,' or go unnamed, as with *gravel*;

---

1. Some positive criterial properties are often recognized, for example, combinability with *much/a lot of, little,* and *some,* but these usually work only for grammatically singular mass nouns.

- whether the noun is compatible with various "detached portion" partitive nouns such as *piece, bit, block, lump,* and *chunk,* and if so, with what selection of them;
- "scale effects": how much of the referent in question counts as *a lot,* for example, 'a lot of snow' and 'a lot of gravel,' which probably refers to a larger quantity than 'a lot of rice' or 'a lot of coffee grounds.'

The distribution of such properties, it must be stressed, is partly language specific. Not all languages have formal analogues of some of these particular properties, for example, singular/plural marking, but even in languages that do have such analogues, their distribution can differ. For example, in Russian the words for *peas* and *beans* (*gorox, gorošek, fasol'*) are mass nouns, just like the words for rice and flour (*ris, muka*). Likewise, the Russian words for *onions* and *potatoes* are also mass nouns (*luk* and *kartoška*). To refer to them individually, derived "singulative" words must be used, for example, *lukovka* or *lukovica* for an individual onion (Wierzbicka 1988).

Some scholars seem reluctant to embrace the notion of a highly variegated nominal lexicon, so it may be helpful to draw an analogy with the verbal lexicon, where recent scholarship has converged on just such a conclusion. When once linguists would have spoken comfortably in terms of a dichotomy between transitive and intransitive verbs, most now accept the existence of dozens of verbal subclasses (for a recent review, see Levin and Rappaport Hovav 2005), which become evident once more subtle syntactic diagnostics are employed than simply the presence or absence of a direct object. The crude "count versus mass" dichotomy among nouns is roughly analogous to the old "intransitive versus transitive" dichotomy among verbs. Both are useful at a first approximation, but if they are reified as absolutes they can become a barrier to progress.

The analogy can be pressed further to questioning the status of the subclasses themselves. After observing that the syntactic behaviors of verbs point to the existence of "semantically coherent classes of verbs," Levin and Rappaport Hovav (2005: 16) remark: "However, other generalizations pick out sets of verbs that are larger, smaller or even partially overlapping with these classes. Such data lead to the conclusion that it is the elements of meaning that define verb classes that are most important, and that verb classes themselves are epiphenomenal." Therefore, they conclude, further advances require "isolating those semantic components which ultimately determine them." One important goal of the present study is to identify the semantic components underlying the existence of some of the language-specific subclasses in the English nominal lexicon. I will continue to speak of subclasses of mass nouns, but this is simply a terminology of convenience.

Inextricably involved in this project—as with any project in conceptual-semantic analysis—is the question of metalanguage. What is the appropriate "semantic vocabulary" for stating the relevant components of lexical semantic representation? This was the other major feature of Wierzbicka (1988: 555): her insistence that the conceptual content of the various subclasses of mass noun could and should be spelt out in "formal semantic representations (explications)." Informal discussion, Wierzbicka stated, was "not good enough"—and if many linguists were satisfied with it, this was because "they are not used to taking semantics seriously: they simply don't expect from semantics the methodological rigor or discipline which in syntax or phonology they would regard as indispensable." The essential tool for formal semantic representation is of course a well-specified semantic metalanguage. Since 1988, NSM, the semantic metalanguage originated by Wierzbicka (1972, 1980), has been greatly improved (Goddard and Wierzbicka 2002; Wierzbicka 1996; see Goddard 1998, 2006a, 2007a). By current standards, the semantic schemas in Wierzbicka (1988) were presented in a relatively "loose" metalanguage. In the present study, the semantic schemas are presented in a much more disciplined metalanguage. My intention is to focus attention on the formal statements of the semantic components underlying count/mass phenomena in English.

Among linguists, the NSM system is sufficiently well known not to require extensive explanation, but it may be helpful to draw attention to its main features. The hallmark of the system is that meanings are represented in a tightly constrained, yet expressively flexible, mini-language of empirically established universal semantic primes along with their inherent universal grammar. Semantic primes are simple indefinable meanings that, evidence suggests, surface as identifiable word-meanings in all languages. These include meanings such as SOMEONE, SOMETHING/THING, PEOPLE, KIND, PART, WHERE/PLACE, WHEN/TIME, DO, HAPPEN, MOVE, TOUCH, THINK, WANT, KNOW, SAY, GOOD, BAD, BEFORE, AFTER, BECAUSE, IF, NOT, and LIKE. The total number of primes currently stands at 63. They are displayed in table 8.1, using exponents from English and Spanish. An extensive bibliography is available at the NSM Homepage (http://www.une.edu.au/bcss/linguistics/nsm/).

The NSM metalanguage aspires to be a formal semantic metalanguage based on natural language. Because its vocabulary and syntax are recruited from ordinary language, the NSM system can achieve much greater clarity and accessibility than is possible with more technical and more obscure modes of representation. Because the metalanguage is minimal in size, it can allow maximum resolution of semantic detail and ward off any possibility of circularity.

Not all lexical meanings can be resolved directly or immediately to the level of semantic primes, however. Some are best explicated in stages, using intermediate-level "semantic molecules." This term refers to complex word-

TABLE 8.1 Semantic primes: English and Spanish exponents

Substantives	I, YOU, SOMEONE, SOMETHING/THING, PEOPLE, BODY	YO, TU, ALGUIEN, ALGO/COSA, GENTE, CUERPO
Relational substantives	KIND, PART	TIPO, PARTE
Determiners	THIS, THE SAME, OTHER/ELSE	ESTO, LO MISMO, OTRO
Quantifiers	ONE, TWO, MUCH/MANY, SOME, ALL	UNO, DOS, MUCHO, ALGUNOS, TODO
Evaluators	GOOD, BAD	BUENO, MALO
Descriptors	BIG, SMALL	GRANDE, PEQUEÑO
Mental predicates	THINK, KNOW, WANT, FEEL, SEE, HEAR	PENSAR, SABER, QUERER, SENTIR, VER, OÍR
Speech	SAY, WORDS, TRUE	DECIR, PALABRAS, VERDAD
Actions, events, movement, contact	DO, HAPPEN, MOVE, TOUCH	HACER, PASAR, MOVERSE, TOCAR
Location, existence, possession, specification	BE (SOMEWHERE), THERE IS, HAVE, BE (SOMEONE/SOMETHING)	ESTAR, HAY, TENER, SER
Life and death	LIVE, DIE	VIVIR, MORIR
Time	WHEN/TIME, NOW, BEFORE, AFTER, A LONG TIME, A SHORT TIME, FOR SOME TIME, MOMENT	CUÁNDO/TIEMPO, AHORA, ANTES, DESPUÉS, MUCHO TIEMPO, POCO TIEMPO, POR UN TIEMPO, MOMENTO
Space	WHERE/PLACE, HERE, ABOVE, BELOW, FAR, NEAR, SIDE, INSIDE	DÓNDE/SITIO, AQUÍ, ARRIBA, DEBAJO, CERCA, LEJOS, LADO, DENTRO
Logical concepts	NOT, MAYBE, CAN, BECAUSE, IF	NO, TAL VEZ, PODER, PORQUE, SI
Intensifier, augmentor	VERY, MORE	MUY, MÁS
Similarity	LIKE	COMO

Primes exist as the meanings of lexical units (not at the level of lexemes). Exponents of primes may be words, bound morphemes, or phrasemes. They can be formally complex. They can have combinatorial variants (allolexes). Each prime has well-specified syntactic (combinatorial) properties.

meanings, themselves decomposable into semantic primes, which appear as units in the explications of yet more complex concepts (see Wierzbicka 1996: 221, 2006a, 2006b, in press; Goddard 1998: 254–255, 2006a, 2007b). For example, body-part meanings such as 'mouth,' 'hands,' and 'legs' function as semantic molecules in the explications for physical activity verbs, such as *eat*, *hit*, and *run*. When semantic molecules are used in explications, they are marked as such by the notation [M].

NSM semantic explications and schemas have to meet three conditions: (i) *well-formedness*, that is, they have to be framed exclusively within NSM:

TABLE 8.2 Semantic schemas and explications in the main body of this chapter

Section 2 Prototypical count nouns

| [A] | *cat/cats, bottle/bottles* [regular count nouns] |

Section 3 Concrete mass nouns

[B]	*cheese, glass, paper* [homogeneous substances]
[C]	*rice, sand, salt, snow* [particulate substances with named minimal units]
[C']	*dust, flour, powder* [particulate substances with minute, named minimal units]
[D]	*gravel, mulch, straw* [particulate substances without named minimal units]
[E]	*oats, chives, split ends* [aggregates of small unnamed things, marginally individualizable]
[E']	*dregs, curds, suds,* [aggregates of small unnamed things, not individualizable]
[F]	*noodles, peas, beads* [plural-mostly nouns: aggregates of small individually named things, conceptually belonging together]

Section 4 "Minimal unit" nouns

[G]	*grain (a grain of rice, sand, salt, ...)*
[H]	*flake (a flake of snow, dandruff)*
[I]	*speck (a speck of dust, flour, talcum powder, ...)*

Section 5 "Detached portion" nouns

[J]	*bit (a bit of butter, paper, cake, ...)*
[K]	*piece (a piece of paper, cheese, glass, ...)*
[L]	*lump (a lump of clay, meat, coal, ...)*
[M]	*block (a block of wood, ice, cheese, ...)*
[N]	*chunk (a chunk of cheese, chocolate, metal, ...)*
[O]	*chip (a chip of stone, wood, ...)*
[P]	*slice (a slice of bread, lemon, salami, ...)*

(ii) *coherence*, that is, they have to make sense as a whole; and (iii) *substitutability* in a broad sense, that is, they have to make intuitive sense when substituted into their contexts of use, generate the appropriate entailments and implications, and so on.

Two further general points perhaps warrant emphasis. First, the NSM system aims to be exhaustive, that is, capable of adequately paraphrasing the entirety of the vocabulary of any language. There is a large body of descriptive-analytical work in the framework, dealing with lexical, grammatical, and illocutionary semantics across many domains and across many languages. In other words, although the present study has a rather narrow focus, the metalanguage it uses has not been specially devised for this purpose, but on the contrary is merely an application of a general metalanguage of semantic explication applicable across the entire lexicon and grammar of any language. Second, the NSM system is intended to represent conceptual reality—not of specialists in logic, philosophy, or mathematics, but the conceptual reality of ordinary language users, as embedded and embodied in their language. As such, it ought to be open to empirical testing using psycholinguistic methods, evidence from language acquisition, and so on, and it is heartening that such research is under way in some quarters (e.g., Middleton et al. 2004; Tien 2005; Wisniewski et al. 1996; Wisniewski et al. 2003). The present study, however, is confined to linguistic evidence.

For convenience, table 8.2 summarizes the semantic schemas and explications of count and mass words that are presented in the main body of this chapter.

2. A Semantic Schema for Prototypical Count Nouns

The "category" of mass nouns exists in contrast with that of so-called count nouns, which exhibit the familiar singular/plural contrast, are compatible with the singular indefinite article *a(n)*, and so on. Without suggesting that every such noun conforms to this schema, I would like to propose schema [A] below as a characterization of the semantic content shared by prototypical count nouns—roughly, what philosophers sometimes term "sortals." There are vast numbers of English nouns that conform to this schema. It is a deeply ingrained semantic pattern of the English language.

[A] **cat/cats, bottle/bottles** (i.e., regular count nouns)
   a. something of one kind
   b. people can say what kind with the word *cat* (*bottle*, etc.)
   c. people can think about something of this kind in one of these ways:
   d.   "this something is one thing of this kind"
        "this something is not one thing of this kind, it is more things of this kind"

As one can see, the schema itself is constructed from plain ordinary words, without recourse to any technical vocabulary such as "sortal," "instantiation," "construal," "individuation," or "countability." Nonetheless, each component can be linked with the essential content of notions such as these. Component (a) indicates that the noun in question represents 'something of one kind,' that is, a sortal. The count noun word itself (in its bare uninflected form) is then presented, in component (b), as a way in which people can identify the kind of thing in question (see Xu's [2005] notion that category names act as "essence placeholders"). Component (c) then specifies that any particular instantiation of the kind in question is subject to alternative construals, that is, 'people can think about something of this kind in one of these ways.' The first possibility is 'this something is one thing of its kind,' that is, a single instance of the kind in question. The other possibility comes in two parts. The first part is the contrary of the first option, that is, 'this something is not one thing of this kind,' which clearly implies the potential for "individuation." The second part continues: 'it is more things of this kind.' Obviously, these alternative construals correspond to the existence of singular and plural forms, respectively.[2]

Needless to say, there are many semantic subclasses of prototypical count nouns, and any individual noun will be specified in much greater detail than in schema [A]. NSM studies of natural kind terms (e.g., *cat*) and artifact terms (e.g., *bottle*) indicate that such words encapsulate an extremely rich and dense assemblage of semantic components (e.g., Wierzbicka 1985, 1996, in press), often running to twenty or more lines of semantic text and incorporating multiple levels of nested semantic molecules. One can think of schema [A] as representing the shared topmost level of a vast number of English noun concepts that otherwise differ greatly in their semantic structure and content.

Though schema [A] does not contain any literal reference to countability as such, the fact that referents falling under this schema can be thought of as 'one thing of this kind' makes them countable in principle (compare Spinoza, Frege). This is because, from a semantic point of view, a basic prerequisite for counting things is that one can think of them "one by one," matching each cognitive act with a word in the counting sequence (*one, two, three, four*, etc.) (Goddard in press). It would be entirely possible to draft a more literal "countability" component in the NSM metalanguage, for example, 'someone can say (with one word) how many things of this kind there are in a place,' and such a component may be warranted in languages that exhibit plural number marking only in the case of literally countable items. For English, however, with its near obligatory number marking, such a component would be too strong. There are simply too many count nouns (e.g., *stars, clouds, mosquitoes*) that are not countable in a practical or literal sense.

3. Seven Subclasses of Concrete Mass Noun

Schema [B] below is intended to capture the shared conceptual content of homogeneous substance nouns, such as *cheese, glass, paper*, and *iron*. The first two components are the same as for other sortal nouns. The distinctive component is component (c), which denies the possibility of thinking of such a something of this kind in terms of multiple enumerable parts.

[B] cheese (glass, paper, butter, coal, iron) (i.e., homogeneous substances)
a. something of one kind
b. people can say what kind with the word *cheese* (*glass, paper, butter*, etc.)
c. people can't think about something of this kind like this:
   "this thing has many parts, someone can know how many"

---

2. In a language with dual marking, there would be an additional option: 'it is two things of this kind,' and the final component would be expanded to exclude both singular and dual options.

Needless to say, the lack of any individuation components (comparable to those found in schema [A]) is correlated with the absence of the singular/plural morphology.

Schema [C] below applies to "particulate substance" nouns such as *rice, wheat, grass, snow, sand,* and *sugar.* These differ from the homogeneous substances just dealt with in that they are conceptualized as consisting of multiple identical "minimal units," such as *grains* or *flakes.* Under component (c) of the following schema it is stated that instantiations of things of this kind can be thought of as 'having many very small parts, all of the same kind.' As we will see, the specification that the parts are 'of the same kind' is linked with the existence of a minimal unit word, such as *grain*, to refer to them. (There are other kinds of particulate substances which lack such minimal unit words, and the distinguishing feature seems to be whether or not the constituent parts are thought of as themselves belonging to a particular kind.)

[C] rice, wheat, sand, sugar, salt, snow, grass (particulate substances with named minimal units, e.g., *grains, flakes, blades*)
a. something of one kind
b. people can say what kind with the word *rice* (*sand, grass*, etc.)
c. people can think about something of this kind like this:
   "this something has many very small parts
   all these very small parts are of the same kind"

There are, of course, important semantic differences between foodstuffs such as *wheat, rice, salt,* and *sugar,* and natural geographical phenomena such as *sand, snow,* and *grass.* Such differences are not directly relevant for present purposes, except to the extent that they help provide some grounds or motivation for differences in the practical countability of instances of the referents in question. For example, though in most circumstances one would hardly imagine anyone being interested in how many grains of *rice* or *sugar* there are in a particular location, it is not altogether inconceivable. A person might notice and remark on 'a couple of grains of rice,' for example, being left on a plate, or 'a few grains of sugar' being spilt on the kitchen bench. The possibility arises because *rice* and *sugar* are used in domestic settings, and in relatively small quantities. On the other hand, it is not so easy to imagine anything comparable happening with geographical referents like *sand* or *grass,* which are normally encountered outdoors and in large quantities.

The situation is different with things like *dust, flour,* and *powder.* They, too, are made up of lots of minimal units (e.g., *specks*), but these are by their nature so very small and so very numerous as to be a lot less perceptually and cognitively salient than *grains*, say. While one is aware of the existence of specks and they are marginally countable (e.g., 'There was a speck or two of dust/powder on the glass'), one does not normally think about individual *specks*. For substances like these, it would seem justifiable to include an additional detail in the (c) component, as per schema [C′] below: 'people cannot see one of these very small parts if they don't do some things at the same time'; that is, they are ordinarily not visible individually without some special effort.[3]

[C′] dust, flour, powder (particulate substances with minute, named minimal units, i.e., *specks*)
a. something of one kind
b. people can say what kind with the word *dust* (*flour, powder*, etc.)
c. people can think about something of this kind like this:
"this something has many very small parts
all these very small parts are of the same kind
people cannot see one of these very small parts
if they don't do some things at the same time"

Yet another variety of concrete mass noun is represented by words such as *gravel, mulch, straw,* and *litter*. Like the previous two groupings, they have a variegated internal composition, but for some reason there are no names that can be used to refer to the small individual items that compose them. As Wierzbicka (1988) remarked, this might seem especially curious in view of the fact that the items in question are usually not very small (small perhaps, but not very small).

The evident reason is that the constituent items are noticeably variable in size and shape. The *Longman Dictionary* (1987: 457), for example, defines *gravel* as "small stones usu. mixed with sand." The internal diversity is increased by the fact that there are different kinds of *gravel, mulch*, and *straw*. For example, in my garden we have two kinds of *mulch*, one made of dark-colored largish wood chips and the other of small pale flake-like chips. Even with *straw*, there can be different thicknesses and lengths of the individual bits.[4] Accordingly, in component (c) of schema [D] there is the information that 'this something has many small parts,' but no specification that these parts are all of a single kind. The lack of any word naming the parts is linked with the fact that they are not seen as constituting a distinctive 'kind' in their own right.

Gravel, mulch, straw, and so on, exhibit what I will call the "large-scale effect"; that is, one tends to think of them occurring in fairly large quantities

---

3. The minute size and virtual "invisibility" of individual *specks* make them uncountable in a practical sense, but it does not seem necessary (or intuitively appealing) to add a literal component to this effect, for example, 'no one can say how many of these many very small parts there are in one place.'

4. Admittedly, there is the expression *to draw the short straw*, but this applies to individual items that have been selected and made comparable to one another.

and spread out over a wide area (see Wierzbicka 1988: 531). I think this effect can be adequately accounted for by the final line of component (c): 'it can be like part of the ground [M] in a place.' 'Ground [M]' is, of course, a semantic molecule, not a semantic prime.

[D] gravel, mulch, straw, leaf litter (particulate substances without named minimal units)
a. something of one kind
b. people can say what kind with the word *gravel* (*mulch, straw*, etc.)
c. people can think about something of this kind like this:
    "this something has many small parts
    it can be like part of the ground [M] in a place"

So far, the various subclasses we have considered have all been in the singular (uninflected) form. There are, however, quite a few groups of English mass nouns that appear, more or less invariably, in the plural form—words that are sometimes termed *pluralia tantum* or plural only. In the case of "dual object" words such as *scissors, pants, glasses*, and *headphones* (which are also compatible with the *pair of* construction), there is a pretty obvious semantic explanation based on the bipartite nature of the referents (or rather, on how the referents are conceptualized) (Wierzbicka 1991). But what about more "stuff-like" kinds of thing, such as *oats, coffee grounds, grass clippings*, and *hundreds-and-thousands* (an Australian word for a cake topping, sometimes called *sprinkles*)?

The example of *oats* has something of a *locus classicus* status, given that it was used by the structuralist linguists of last century to argue for the arbitrariness of the morphology. What possible semantic motivation could there be, it was asked, for *oats* to carry an invariable plural morphology, when other apparently similar referents such as *wheat* (*rice, barley*, and others) appear in uninflected form? To muddy the waters further, some plural-only words do not seem particularly "stuff-like" either—for example, *chives, split ends, bean sprouts, suds*, and *pine needles*.

To approach this question, it is useful to review the formal linguistic properties in a little more detail. According to Wierzbicka (1988), the words in question all carry a plural -*s* and take plural verb agreement, but they cannot appear in the singular form or be quantified with a numeral (*an oat, *three oats). It seems to me, however, that some of these words are marginally acceptable with vague numerical quantifiers, such as *a couple* and *a few*, for example, *a couple of chives, a few oats*. Others, however, are quite out of the question, for example, **a couple of dregs*, *a few curds. Likewise, some of the words (or rather, their referents) can be individuated to a limited extent, for example, 'I was bringing in some chives from the garden and I dropped one on the kitchen floor,' while others are completely resistant to any such treatment. I would go so far as to say that in some cases, the singular form can be pressed into service when there is a need to refer to an individual item, e.g., ?*an oat*, ?*a split end*, ?*a bean sprout*, while in other cases such a usage is completely beyond the pale, for example, *a dreg, *a sud, *a curd, *a dropping. On this evidence, I think it is necessary to recognize two subclasses of these plural-only words.

Schema [E] applies to what seems to be the larger of the two classes, which tolerates "small" vague quantifiers and limited individuation, and sometimes allows a singular usage. This schema is very different from the others proposed thus far, inasmuch as it is not presented in terms of a parthood structure. Component (c) does not say that something of this kind 'has many (very) small parts.' Rather, it says that something of this kind is composed of multiple little things—using the semantic prime "BE of specification." The claim is, in other words, that whereas referents such as *rice, sand, grass*, and *dust* are thought of as "having lots of little parts," referents such as *oats, split ends*, and *bean sprouts* are thought of as "*being* lots of little things." The presence of plural marking is of course the formal morphological correlate of this semantic component. The next line says in effect that many of these small items can be together 'in one small place.' This component also helps account for the "small-scale effect" noted by Wierzbicka (1988: 531), in connection with words in this grouping. The final line of component (c) goes on to further characterize instances of this kind of thing by way of a new component: 'someone can do some things to one of these things with one finger [M].' This serves at once to indicate size and to suggest a degree of potential individuation, inasmuch as it envisages some minimal individual handling of the items. This tends to imply that they are more concrete and separable than their close cousins, such as *dregs* and *coffee grounds*, to be considered next. The component is not phrased in terms of 'holding' the item for two reasons: first, it is not 'holding' as such that is the relevant property, but something more like manipulability; and second, 'holding' would not be applicable to *eye drops* in any case.[5]

[E] oats, chives, split ends, eye drops, bean sprouts, pine needles, hundreds-and-thousands, curls, potato peels, grass clippings, tea leaves, iron filings, shavings (aggregates of small unnamed things, marginally individualizable)
a. something of one kind
b. people can say what kind with the word *oats* (*bean sprouts, curls*, etc.)
c. people can think about something of this kind like this:
  "this something is many small things
  there can be many of these small things in one small place
  someone can do some things to one of these things with one finger [M]"

---

5. Notice the remarkable presence of the word *eye drops*. This actually refers to a liquid, but the fact that it will be portioned out in individual doses allows the conceptualization that it consists of 'many small things.'

Schema [E′] applies to a smaller group of plural-only words, which as mentioned above, are incompatible even with vague quantifiers, and for which a singular form is absolutely out of the question. Most of the explication is the same as for [E], that is, the characterization of something of this kind as consisting of 'many small things' and the "small-scale" specification (compare *suds* with *froth*). But the [E′] grouping lacks the final specification with its implications of potential minimal manipulability.

[E′] dregs, curds, suds, droppings, tailings (aggregates of small unnamed things, not individualizable)
 a. something of one kind
 b. people can say what kind with the word *dregs* (*coffee grounds, curds,* etc.)
 c. people can think about something of this kind like this:
  "this something is many small things
  there can be many of these small things in one small place"

A rather different subclass of plural mass nouns, termed "mostly plural" by Wierzbicka (1988), is represented by words like *noodles, peas, beans,* and *grapes*. I will call them "plural-mostly," for terminological parallelism with the "plural-only" nouns considered in the preceding section. The most obvious difference between the two categories is, of course, that the plural-mostly nouns have normal singular forms; and furthermore, it is perfectly possible to count the individual items. For practical reasons people normally wouldn't be bothered to count them in any numbers, but it is perfectly possible to say, for example, that there are *two noodles* or *three peas* left on a plate.

On the other hand, a word like *noodles, peas, beans,* and *grapes* is far from a regular count noun, such as would fall under schema [B]. The point is that it is integral to the concept of *noodles* or *peas,* for example, that the kind in question consists of multiple small identical items ('it is many small things, they are all of the same kind'), whereas there is nothing similar in the conceptual content of regular count nouns, such as *cat* or *bottle*.

Schema [F] therefore shares with the two schemas just presented the "multiple things" line in component (c): 'it is many small things.' Unlike as with *oats* and *dregs,* however, this line is followed by the statement that 'these many small things are all of the same kind.' The fact that the individual items are conceptualized as belonging to a single kind is, as usual, associated with the existence of a word for the individual items. The schema also includes a manipulability component, this time referring to the possibility of doing something to the items 'with two fingers [M]'.

Another formal indication of the difference between these nouns and regular count nouns is how they work in relation to a rule of regular polysemy (see the appendix), which derives "mass" senses of words for edible kinds of things. Briefly, regular count nouns of the right kind can be converted to mass nouns by simply stripping them of the plural

TABLE 8.3   Main properties distinguishing the noun subclasses treated in this chapter

	Singular form?	Plural form?	Detached portion? (X of ...)	Named minimal unit?
[B] *cheese, glass, paper*	+	−	+	−
[C] *rice, sand, snow*	+	−	−	+
[D] *gravel, mulch, straw*	+	−	−	−
[E] *oats, chives, split ends*	−	+	−	N/A
[F] *noodles, peas, beads*	+	+	−	N/A

morphology; for example, a sentence like 'I can't eat tomato or mushroom' presents *tomato* and *mushroom* (in the singular form) as foodstuffs or ingredients, whereas 'I can't eat tomatoes or mushrooms' presents *tomatoes* and *mushrooms* as ordinary plurals. The plural-mostly nouns under discussion in this section, however, are not eligible for this conversion, for example, '?I can't eat bean,' '?I can't eat noodle.' In a real sense, the "plural" *-s* is part of the name of the kind of thing in question.

(A cautionary note: as mentioned in the introduction, there are many other subclasses of plural mass nouns, such as *groceries, leftovers, remains, spoils, ruins, odds,* and *earnings*. These each require separate treatment.)

[F] noodles, peas, beans, grapes, beads, eyelashes, tassels, sultanas, lollies (plural-mostly nouns: aggregates of small individually named things, conceptually belonging together)
a. something of one kind
b. people can say what kind with the word *noodles* (*peas*, etc.)
c. people can think about something of this kind like this:
   "it is many small things
   these many things are all of the same kind
   someone can do some things to one of these things with two fingers [M]"

This has not been an exhaustive treatment of concrete mass nouns in English, but it should be sufficient to illustrate that there are multiple subclasses, each of which can be assigned a distinctive semantic schema. It should also be clear that to a large extent the formal properties of these subclasses make sense in terms of the semantic schemas. For convenience, the major subclasses and the formal properties that distinguish them are summarized in table 8.3 (a number of minor properties and subclasses mentioned in the text have not been included).

4. Minimal Unit Words: *Grain, Flake, Blade, Speck*

Let us now consider minimal unit nouns, such as *grain, flake, blade,* and *speck*. All these words are in a sense prefigured by the general schema proposed in section 2 for aggregate mass nouns such as *rice, wheat, sand,* and *sugar*. For convenience, this schema is repeated below. A slightly more elaborate variant [C'] was adduced for such nouns as *dust, powder,* and *flour,* whose minimal parts are tiny and virtually invisible on an individual basis (see section 4.3 below).

> **[C] rice, wheat, sand, sugar, salt, snow, grass** (particulate substances with named minimal units, e.g., *grains, flakes, blades*)
> a. something of one kind
> b. people can say what kind with the word *rice* (*sand, snow, grass,* etc.)
> c. people can think about something of this kind like this:
>   "this something has many very small parts
>   all these very small parts are of the same kind"

Of all the minimal unit words, perhaps the most interesting—and certainly the most common—is *grain*. One of the most interesting things about it, shared to a limited extent by *flake*, is that it can be used about things that might appear to be very different in their basic nature. Certainly from the point of view of physical reality, three different cases can be distinguished. A *grain of wheat* (or *rice*) is a natural part of the biological world. *Grains of wheat, rice,* and so on, are in fact seeds (though they are not being conceptualized as such when referred to as *grains*), and they therefore have a certain stability of shape and size, and also a certain "integrity." *A grain of wheat* (or *rice*), for example, refers to a whole thing: half *a grain of wheat* (or *rice*) is not a *grain* at all. In contrast, *a grain of sugar* (or *salt*) is not a biological kind of thing, and *grains of sugar* (*salt,* etc.) can be expected to vary quite a bit in shape and size. They also lack the integrity of *grains of wheat* or *grains of rice*. After dividing a biggish *grain of sugar,* for example, one could well be left with two things that could each qualify as *a grain of sugar* in its own right. *Sugar,* moreover, is a domestic product, albeit it is extracted from plants like sugarcane and sugar beet. One could see *sand,* and *grains of sand,* as intermediate between these two cases. *Sand* is presumably thought of as part of the natural world, as a natural phenomenon, so to speak, and in this sense *grains of sand* are more natural than *grains of sugar*. On the other hand, one knows that *grains of sand* can vary significantly in size and/or texture, and they are divisible.

Not surprisingly, some languages have different words for these different items. In Polish, for example, different words are used for 'a grain of wheat' (*ziarno*) and 'a grain of sand/salt' (*ziarnko*). The latter is morphologically derived from the former by means of a specialized diminutive formation. (There is also a general diminutive form of the word for 'a grain of wheat, etc.' (*ziarenko*), but this cannot be used in reference to a grain of sand.)

The question is—do these differences mean that the English word *grain* is polysemous? In my view, there is no compelling reason to think so. From a linguistic and conceptual point of view, the English concept of *a grain* seems to gloss over and ignore the real-world differences just described. The meaning of an English phrase like 'a grain of wheat' or a 'grain of sand' can be explicated as in [G] below. The top component (a) reads simply: 'something very small.' The claim here is that from a semantic point of view, words like *grain* (and other partitive nouns, including *piece, lump*, etc.) are not sortals: a *grain* is not a "kind" of thing, even though *a grain of sand* or *a grain of wheat* is.[6] The components in (b) are descriptive: they characterize a *grain* as 'something hard [M]' and state that 'someone can do some things to one of these things with two fingers [M].' As before, this component serves to indicate size from an anthropocentric point of view and, at the same time, suggests the possibility of some human interaction with individual items. The most distinctive component of the explication is (c): 'these things are all very small parts of something of one kind.' Component (d) goes on to identify the kind of thing as *wheat, rice, sand, sugar*, or whatever, as appropriate.

[G] a grain of wheat (rice, sand, sugar, ...)
a. something very small
b. it is something hard [M]
someone can do some things to one of these things with two fingers [M]
c. these things are all very small parts of something of one kind
d. people can say what kind with the word *wheat* (*rice, sand, sugar*, etc.)

Flakes *and* Specks

The situation with *flake* at first seems to parallel that of *grain*, at least in part. There are a few naturally occurring kinds of *flake*—*flakes of snow* (*snowflakes*), *flakes of bran, flakes of dandruff*. On the other hand, there are a range of other uses with a variety of referents, for example, *flakes of skin, flakes of soap*, and (in archeological parlance) *flakes of quartz* (*quartz flakes*) and *flakes of stone* (*stone flakes*), referring to certain kinds of stone tools. These other uses, however, are not parallel to *grains of sugar* or *grains of salt*, because they all imply that the item in question was previously part of, or at least attached to, something else; *a flake of skin*, for example, was previously part of someone's skin. Furthermore, in these uses, there seems to be what I call

---

6. Since *grains* (*flakes, specks*, etc.) are themselves count nouns, it can be asked whether they ought not include the normal "individuation" components attributed to prototypical count nouns in schema [A]. Briefly, I do not think that these particular components are a *sine qua non* for countability. There are other kinds of countable nouns, for example, body parts and kin terms, which are not sortals.

a "surface effect," that is, the implication that the *flake* was previously part of the surface of the original thing, for example, the expression *to flake off*.

There is another big difference between *grains* and *flakes*. In the case of *wheat, rice, sand*, and even *salt* and *sugar*, it is an integral part of the conceptualization of these substances that they normally consist of numerous very small identical parts, namely, *grains*. In the case of *snow*, I would argue that something similar applies. The idea that it comes in lots of tiny little bits, that is, *snowflakes*, is an integral part of the meaning of *snow* (necessary, among other reasons, to distinguish it from the meaning of *ice*, and to link it with the concept of *rain*). But it is not a necessary part of the meanings of *skin* or *soap* that these substances can sometimes be rendered into *flakes*.

In the case of *flake*, therefore, I think that we do have to recognize polysemy. *Flake*$_1$ is a natural unit of a larger "kind" (e.g., *snow*), whereas *flake*$_2$ is a partitive noun of the "detached portion" variety, like those dealt with in section 3. The schema below applies only to *flake*$_1$. This explication, too, is intended to be "read against" the general mass noun schema [C], for kinds of particulate substances with minimal named units. The descriptive components in (b) say that *a flake*$_1$ is 'thin,' 'flat,' and (roughly speaking) "hard to see"; that is, one can't see them unless one does some other things at the same time (in essence, a special effort is required).

[H] a flake$_1$ of snow (bran, dandruff)
   a. something very small
   b. it is something thin [M], it is something flat [M]
      people can't see one of these things if they don't do some things at the same time
   c. these things are all very small parts of something of one kind
   d. people can say what kind with the word *snow*

I will not undertake an explication here for *a blade of grass*. In principle it seems little different from *flake*$_1$, except for being even more specialized in application.[7] Presumably, its descriptive components would include 'it is something long [M]; it is something flat [M]; it has straight [M] edges [M].' There is an obvious parallel with the *blade* of a knife, sword, paper cutter, and so on. Presumably, *a blade of grass* would also include the component 'someone can do some things to one of these things with two fingers [M].'

The word *speck* differs from *flake* and *blade* in that it is not confined to a narrow range of phrase-mates, but co-occurs freely with *dust, powder, flour*, and *pollen*. In this respect, it is more analogous to *grain*. But in terms of semantic content, the main difficulty with *speck* is to capture the aspect connected with the extremely small size, nonmanipulability, and near

---

7. According to some dictionaries, one can speak of *a blade of wheat, blade of barley, blade of rye*, and so on, but this usage is archaic (though preserved in Biblical English).

invisibility, and in this respect it more resembles *flake*. I would tentatively propose the explication in [I].

[I] a speck of dust (flour, talcum powder, pollen ...)
a. something very small
b. people cannot see one of these very small things,
  if they do not do some things at the same time
  someone can't do anything to one of these things with two fingers [M]
c. these things are all very small parts of something of one kind
d. people can say what kind with the word *dust* (*flour, powder*, etc.)

A possible alternative to the "barely visible" component in (b) might be 'because things of this kind are very small, someone cannot see one of these things if it is not very near this someone's eyes [M].'

5. "Detached Portions": *Pieces, Bits, Blocks, Lumps, Chunks, Chips,* and *Slices*

Nouns for detached portions (Climent 1996) differ in number and in semantic content across languages and constitute a fascinating descriptive challenge in their own right. What is the precise difference between *a piece of cheese, a bit of cheese, a block of cheese*, and *a lump of cheese*? What is the difference between a *chunk of chocolate* and *a hunk of chocolate*? Size and shape both seem to be involved somehow, but exactly how?

Detached portion nouns can be regarded as a subclass of "unitizer" words because they allow mass things to be individuated. This property is modeled by component (a) in each of the following schemas, which specifies each as 'one something of one kind,' the kind in question being specified by the noun in the accompanying *of* phrase. *A bit of wood*, for example, is something of the kind *wood; a lump of clay* is something of the kind *clay*, and so on. This aspect is modeled in component (b) in each of the following schemas: 'people can say what kind with the word *wood* (*clay*, etc.).'

As the term "detached portion" suggests, such nouns refer to something that was previously part of something else. A *piece* of something, for example, must have been cut, torn, broken off, or otherwise separated from a larger thing.[8] For this reason, an ice cube that has been frozen whole is not *a piece of ice*. Likewise, an amount of glass that has solidified on the floor of a glassblower's workshop is not *a piece of glass*. Accordingly, a component is needed that refers or alludes to some kind of separation event: 'before, it was part of something else of the same kind; it is not part of this thing anymore because something happened to it.'

As for the differences between the different detached portion nouns, these can be largely captured by way of descriptive specifications about

---

8. Exceptions would be *a piece of gum* (from a package) and *a piece of candy* (from a box of candies).

the form of the referent. For example, for *bit*, 'it is something small'; for *block*, 'it is something hard [M]'; for *slice*, 'it is something flat [M].' As these examples suggest, these components typically employ semantic molecules for shape and dimensional concepts, for example, 'long,' 'straight,' and 'flat,' and for physical qualities, for example, 'hard,' 'soft,' and 'sharp' (see Goddard and Wierzbicka 2007; Wierzbicka 2006a).

Bits *and* Pieces

Explication [J] is for the *bit of* X construction, where X is a concrete mass noun.[9] Aside from the shared components just explained, it has a single, very simple descriptive component (c): 'this thing is something small.' (One of the most common collocations is the expression *little bit/ little bits*; for this, I would posit the component 'this thing is something very small.')

[J] a bit of butter (paper, cake, wood, ...)
 a. one something of one kind
 b. people can say what kind with the word *butter* (*paper, cake,* etc.)
 c. this thing is something small
 d. before, it was part of something else of the same kind
   it is not part of this other thing anymore because something happened to it

Compare explication [J] for *bit* with explication [K] below for *piece*. The word *piece* is interesting because it would seem to be the most versatile and nonspecific of the terms under discussion. Like *bit*, it occurs with a wide range of complement nouns, for example, *a piece of paper, a piece of cloth, a piece of chocolate, a piece of cheese, a piece of thread, a piece of wood, a piece of ice, a piece of glass*, and it does not seem to carry any expectation in terms of size, shape, or other physical property.

It might be thought that something like solidity or rigidity is required, because an expression like **a piece of jam* is completely unacceptable, while *?a piece of butter* sounds mildly questionable (unless the butter is very cold, and so is quite solid). However, *cloth* and *thread* are not at all rigid, and there is nothing wrong with *a piece of cloth* and *a piece of thread*. Expressions like *a piece of putty* and *a piece of clay* are also quite normal. The (c) component is therefore left blank in explication [J]. To explain the unacceptability of **a piece of jam*, I think we have to appeal not to size or shape as such, but to something more like "discreteness." Notice that the first line of component (d) in schema [J] reads not simply 'before, this thing was part of something else of the same kind,' but rather 'before, this thing was *one part* of

---

 9. *Bit* and *a bit* are both very common as "minimizer" modifiers, in expressions like *a bit tired* and *a bit of a problem*. These also involve the concept of 'something small,' but in a figurative mode. *I'm a bit tired*, for example, conveys the message that being tired is no big deal for me. In semantic primes: 'I think about it like this: it is something small' (see Goddard 2006b).

something else of the same kind.' The difference in phrasing is small, but significant, because it conveys the impression that what is now *a piece* was important enough to be seen as 'one part' of the overall thing. A portion of jam does not match this requirement, it seems to me, because *jam* does not have enough internal structure to allow the necessary conceptualization.

The component will also help explain why, when we have lots and lots of extremely small bits of, say, wood or paper, one would be reluctant to refer to them as *?pieces of wood* or *?pieces of paper*. It might seem at first that the problem is one of size (i.e., that they are too small), except that things of extremely small size can readily be referred to as *pieces* in other contexts, and there is nothing contradictory about speaking of *a tiny piece*, for example, 'A tiny piece of glass lodged in his eye.' It is not size as such that matters, it seems to me, but something more like the "significance" of a portion in relation to the original thing.

[K] a piece of paper (cloth, cheese, chocolate, thread, wood, ice, glass, cake):
a. one something of one kind
b. people can say what kind with the word *paper* (*cheese*, etc.)
c. —
d. before, it was one part of something else of the same kind
   it is not one part of this other thing anymore because something
       happened to it

An added attraction of the 'one part' component (or subcomponent) is that it establishes a link with some of the other meanings of *piece*, for example, where it refers to a discrete individual part, for example, *a piece of the engine* (see Cruse 1986: 157–159[10]), or where it refers to 'something of one of these kinds,' as in expressions for an individual item of a particular functional kind, for example, *a piece of jewelry, a piece of luggage*.

Lumps, Blocks, *and* Chunks

One of the key properties of *a lump* is something like shapelessness. This property (or nonproperty) is modeled in component (c) in explication [L] as: 'this thing isn't something flat [M], it isn't something round [M], it isn't something long [M].' That is to say, it is nondescript in terms of shape.

---

10. Cruse (1986: Ch. 7) makes a number of useful observations about 'parts and pieces,' but his discussion goes astray in two ways: first, he does not distinguish between the 'part of' relation per se and its individuated or quantified manifestation, for example, in the phrase *a part of*, which effectively equals 'one part of'; second, he does not pay any attention to the English-specific status of the concept lexicalized in the word *piece*, as compared with the universal status of the 'part of' concept.

[L] a lump of clay (meat, wax, coal, ice, ...):
a. one something of one kind
b. people can say what kind with the word *clay* (*meat*, etc.)
c. this thing isn't something flat [M], it isn't something round [M], it isn't something long [M]
d. before, it was part of something else of the same kind
   it is not part of this other thing anymore because something happened to it

It would seem clear that the expression *a lump of sugar* (for sugar cubes, e.g., 'One lump or two?') is a separate lexicalized meaning. Likewise, when we speak of a *lump* in one's breast or shoulder (a possible manifestation of cancer, or a cyst), this too is a different meaning, if only because the lump in question is not a lump *of* anything (*a lump of cancer, *a lump of cyst).

A block is in many ways the antithesis of *a lump*. Whereas *a lump* is more or less shapeless, *a block* seems to require a "squarish" shape, that is, straight edges and flat sides. *A block* must also must be pretty big; expressions like *?a small block of ice* sound peculiar. I would suggest also that *a block* must be composed of something 'hard.' Certainly the most common collocations are *a block of wood* and *a block of ice*. It is not too uncommon to see the expression *a block of cheese*, but in most cases it is clear in context that the cheese in question is fairly hard (often straight out of the refrigerator). One could hardly have *a block* of very soft cheese, such as camembert or ricotta, because the cheese would not be able to maintain its shape. Given this, one might question the need for the component 'it is something hard [M].' Isn't it sufficiently implied by the need for the thing to maintain the requisite 'straight edges' and 'flat sides'? In most cases, perhaps. But there are some 'soft' kinds of things which could maintain their shape, but are nonetheless impossible with *block*, for example, *a block of cake.

[M] a block of wood (ice, cheese, ...):
a. one something of one kind
b. people can say what kind with the word *wood* (*ice, cheese*, etc.)
c. this thing is something big, it is something hard [M], it has straight [M] edges [M], it has flat [M] sides [M]
d. before, it was part of something else of the same kind
   it is not part of this other thing anymore because something happened to it

Like *a block*, *a chunk* must be pretty big (e.g., *?a small chunk*). It can in fact be very big, for example, *a chunk of space rock, a chunk of an iceberg*. The term *chunk* can be usefully compared with *piece*, because, roughly speaking, *a chunk* is *a big piece*. This suggests that component (d) in the explication for *chunk* should perhaps also be specified as 'one part' (rather than simply 'part') of the original item.[11]

[N] a chunk of cheese (chocolate, meat, metal, ...):
a. one something of one kind
b. people can say what kind with the word *cheese* (*chocolate, meat, debris,* etc.)
c. this thing is something big, it is something hard [M]
d. before, it was one part of something else of the same kind
   it is not one part of this other thing anymore because something
      happened to it

Some kinds of peanut butter can be described as *chunky*, but each piece of peanut does not really count as *a chunk* if looked at separately (and, of course, the wording is *chunky*, i.e., 'like chunks'). Likewise, if we speak of a soup 'with big chunky pieces of potato' in it, the individual pieces can't really be spoken of as *chunks*.[12]

Chips *and* Slices

Prototypically, perhaps, *chips* seems to refer to wood that has been chopped,[13] but the word *chip* can apply readily to other results of something hard being hit with something else hard; for example, a piece of *stone* or *ice* can be "chipped" by striking it with a hard instrument. The explication in [O] characterizes *chips* as 'very small' and as being 'something hard [M].'

[O] a chip of stone (wood, ...):
a. one something of one kind
b. people can say what kind of thing with the word *stone* (*wood,* etc.)
c. this thing is something very small, it is something hard [M]
d. before, it was part of something else of the same kind
   it is not part of this other thing anymore because something
      happened to it

Clearly, the expressions *chocolate chips* and *potato chips* are lexicalized names for certain kinds of prepared foods. They are (or are supposed to be) like *chips* in shape, but one could not refer to such food items as **chips of chocolate* or **chips of potato*.

---

11. There are a lot of computer-related uses of *chunk*, for example, *a chunk of text, a chunk of memory*. These represent a separate meaning, but the component 'one part' is no doubt present and constitutes a link with the meaning under consideration. In addition, in the parlance of computing there is the verb *to chunk*, which refers to the grouping together of a number of items so that they function as 'one part' of a larger whole.
12. In view of its size component, the word *chunk* is very suited to figurative uses in the service of expressive "exaggeration," as in *I got a chunk of mascara in my eye. It was terrible.*
13. According to the *OED*, there is no etymological relationship between *chip* ~ *chop*.

Coming now to *slice*, many uses (e.g., *a slice of bread, a slice of salami*) are connected with eating, but we can equally well cut some *slices* of a sample when preparing a slide for a microscope, for example.

[P] a slice of bread (lemon, salami, etc.):
a. one something of one kind
b. people can say what kind with the word *bread* (*lemon, salami,* etc.)
c. this thing is something flat [M], it is something thin [M]
d. before, it was part of something else of the same kind
   it is not part of this other thing anymore because something happened to it

(The expressions *a slice of pie* and *a slice of cake* do not fit explication [P], because the things they refer to are not necessarily thin. I am inclined to think that these are lexicalized expressions, expressing a different (albeit related) meaning to that which occurs in *a slice of bread, a slice of lemon*, etc. Alternatively, one could consider removing the specification 'it is something thin [M]' from the explication.)

There are quite a number of detached portion nouns, such as *splinter, sliver*, and *hunk*, which I have not mentioned in this section, but it seems reasonable to expect that the schema developed here would be adequate for dealing with them.

6. Discussion

Though there are many other mass (i.e., noncount) noun categories that have not been considered in this study, we have covered enough ground to compare the analytical approach adopted here with other approaches in linguistics and in the philosophy of language, and to draw out some general conclusions.

Conventional approaches tend to describe mass/count phenomena in terms of a pair of dichotomies or dimensions, which may be regarded as semantic or syntactic (or both). For example, Jackendoff (1992) uses "±bounded" and "±internal structure"; Gillon (1992) uses ±CT (count) and ±PL (plural), Payne and Huddleston (2002) use "boundedness" and "countability." As shown in the present study, neat dichotomous analyses wrought from materials such as these are not adequate to the facts. They are unable to provide sufficient granularity to account for the multiplicity of subclasses that are identifiable on formal grounds. This failing is not evident in the works of the authors just mentioned, because they typically ignore or dismiss the existence of plural-only and plural-mostly classes and/or ignore the difference between particulate substances with named units as opposed to those without named units, and so on. In other words, they simply close their eyes to data that do not fit the neat binary oppositions favored by their theoretical apparatus. The much richer analyses presented in the present study also achieve greater explanatory force than simpler analyses. The formal linguistic properties of the various subclasses can be seen to be motivated, in considerable detail, by the semantics.

To make these points more concrete, it is useful to compare the NSM treatment of mass nouns and partitive nouns with the approach taken by Ray Jackendoff (1990, 1992, 2002) in his Conceptual Semantics framework. Jackendoff's framework is a pertinent point of comparison because it shares some important meta-assumptions with the NSM approach, while differing on others. The primary commonalities are that the NSM approach and Jackendoff's Conceptual Semantics are both explicitly conceptual in orientation, and that they both accept the existence of universal fundamental conceptual units, which ought to be discoverable by detailed semantic analysis. The main theoretical difference is that Jackendoff's system is an "abstract" one, whose terms and principles of combination bear no direct relation to the words and grammar of ordinary languages. Consequently, Jackendoff's system is much less constrained than the NSM system, and his analyses are for the most part completely opaque to the untrained user.

Jackendoff's (1992) study traverses a number of issues concerned with mass nouns, partitive nouns, and related issues. The key to his whole analysis is a pair of "fundamental conceptual features" $\pm b$ and $\pm i$. The former is linked with the property of "distributed reference"; that is, an unbounded referent can be divided into smaller parts that still qualify as instances of the referent in question, whereas bounded ones cannot (e.g., *water* can be arbitrarily divided into smaller portions of *water*, whereas a comparable result cannot be obtained by dividing up a *pig* or a *bus*). Bare mass nouns and plurals are both regarded as $-b$. The difference between them is that "plurals entail a medium comprising of multiplicity of distinguishable individuals, whereas mass nouns carry no such entailment" (1992: 19). Jackendoff refers to this difference as one of "internal structure" and designates it by way of the feature $-i$. Applied to objects and substances, the feature system comes out like this:[14]

$+b, -i$: individuals(*a pig*) $+b, +i$: groups(*a committee*)
$-b, -i$: substances(*water*) $-b, +i$: aggregates(*buses, cattle*)[15]

Entities of all four kinds are regarded as belonging to a single supercategory "Material Entity" (Mat for short). Thus, individual things receive the analysis [Mat, $+b$, $-i$] and substances [Mat, $-b$, $-i$]. According to Jackendoff,

---

14. In line with his conceptualist orientation, Jackendoff (1992) does not intend the features b or i to designate objective properties, but rather to indicate the speaker's point of view. For example, $-b$ indicates that "boundaries are outside the current field of view ... that we can't see the boundaries from the present vantage point" (19). Likewise, $-i$ does not mean a lack of internal structure in an absolute sense, but rather "lack of necessary entailment about internal structure" (20).

15. Jackendoff (1992: 20) uses the term "aggregates" to refer to "entities normally expressed by plural nouns."

only individual things "have an inherent shape... it is the only category that has physical boundaries. (Groups are bounded in quantity but do not have an inherent shape)" (1992: 20).

The relationship between singular and plural instances of a given noun, and between count and mass instances of a given noun, are dealt with by "conceptual functions" that map between values of b and i. The plural form *dogs*, for example, is obtained by the operation of a function PL that maps a lexical entry for *dog* (with values +b, −i) onto a more elaborate aggregate structure with values −b, −i. (The base element *dog* is retained, embedded within the larger structure.)

Expressions like 'a grain of rice' or 'a stick of spaghetti' are dealt with by way of another conceptual function ELT ('element of'). About the grammatical mass noun *rice*, Jackendoff (1992: 22) says without further explanation that "it happens that it denotes an aggregate rather than a substance"; therefore, to represent the meaning of *a grain of rice* "we need a function that maps an aggregate into a single element of the aggregate."

$$a\ grain\ of\ rice = \begin{bmatrix} +b, -i \\ \text{Mat} \quad \text{ELT}\left(\begin{bmatrix} -b, +i \\ \text{Mat} \quad \text{RICE} \end{bmatrix}\right) \end{bmatrix}$$

The function ELT is described as "an extracting function: the function maps its argument into a subentity of the larger entity denoted by the argument" (Jackendoff 1992: 23). Jackendoff recognizes a second extracting function as GR ("grinder"), whose "argument is an individual or group, and which maps into a substance or aggregate from which the individual or group is composed" (26). For example, in the case of 'dog all over the road':

$$dog\ (\text{substance}) = \begin{bmatrix} -b, -i \\ \text{Mat} \quad \text{GR}\left(\begin{bmatrix} +b, -i \\ \text{Mat} \quad \text{DOG} \end{bmatrix}\right) \end{bmatrix}$$

Climent (1996) extends Jackendoff's system to accommodate more detail about various kinds of partitive nouns, such as *slices, lumps,* and *grains* (among others). He does this by adopting aspects of Pustejovsky's (1995) theory of qualia, but the technical details need not concern us here. For our purposes, the main point is that, as we would expect, it is necessary to augment Jackendoff's very schematic system with provisions for quantity, shape, consistency, thickness, and "process of bringing about," among other things.

Having reviewed these details of Jackendoff's (1992) and Climent's (1996) treatments, we are in a position to make a number of points. First, despite the claim that the terms of the analyses represent "fundamental conceptual notions," it is evident that they cannot be regarded as conceptually simple. We can see this from the way that Jackendoff feels free to add explanatory glosses and explanations employing a much richer vocabulary, such as "individuals," "multiplicity," "element," "subentity," and "shape." To put the point in another way, representations such as those illustrated above are not semiotically self-contained: they require expansion and explanation in a variety of nonobvious ways. In this sense, they cannot be regarded either as fully explicit or as fully resolved, and nor are they as compact as they look.[16] In contrast, the NSM schemas and explications presented in the present study are intended to be semiotically self-contained and fully explicit and to employ only maximally simple constituents.

Second, Jackendoff's (1992) system, and Climent's (1996) extension of it, often obscure, or at least draw attention away from, semantic coherences and explanations. For example, there is evidently some conceptual affinity or compatibility between the notion of an "aggregate" expression and the function "element of," ELT (such that ELT can operate on aggregate expressions such as *rice* and *spaghetti* to produce expressions such as 'a grain of rice' or 'a stick of spaghetti'), but Jackendoff does not explain or comment on this affinity, nor does he explain why it is that ELT can operate on some aggregates, but not on others (e.g., *gravel*, *mulch*, and *litter*).

Perhaps more significant in the bigger picture, however, is the theoretical question of whether an abstract and opaque metalanguage is really suitable for the purpose of conceptual representation. This question is pertinent not only to Jackendoff's Conceptual Semantics but also to any system that purports to represent the conceptual content of ordinary language as used by ordinary speakers. In my view, any such system ought to feel obliged to provide some mechanism whereby its representations can be "translated" into a form that is intuitively accessible to ordinary language users themselves and that generates testable predictions about ordinary linguistic usage. Pelletier and Schubert (2003: 304) make a similar point when they note that formal semantic analyses of mass expressions "are [often] accompanied by informal (i.e., English) paraphrases or elucidations that indicate how these constituents are to be understood intuitively." "We should take such paraphrases seriously," they say, and hence require that "substitution of alleged paraphrases for the original phrases should map from one intelligible sentence into another intelligible sentence synonymous with the original—apart, perhaps, from pragmatic defects such as unwanted implicatures."[17]

According to the present study (and a similar result has emerged from numerous other NSM studies), it is expecting too much for a fully resolved

---

16. As Levin and Rappaport Hovav (2005: 74) point out, it is far from clear how many fundamental notions and functions are at play at any one time in Jackendoff's system, because additional primitive predicates and diacritics are allowed to proliferate freely and with little discussion of their interrelationships.

17. Pelletier and Schubert (2003: 304) also require that "[i]t should be possible to analyse the paraphrase formally, showing that it leads to a translation logically equivalent to the translation of the original."

paraphrase to be expressible as one single intelligible sentence. A sequence of intelligible sentences will normally be called for—a "semantic text," rather than a single sentence—but fully intelligible and capable of being understood intuitively by the ordinary language users whose conceptualization it is supposed to represent.

In contemporary linguistic semantics, the NSM program originated and led by Anna Wierzbicka is the only program that has devoted sustained effort toward satisfying this requirement and that has, furthermore, tested and refined its emerging results in a broad range of semantic domains and across a wide range of languages.

Returning to the subject matter of mass nouns and unitizers, the general conclusions are as follows. First, the nominal lexicon of English, even restricting ourselves to concrete nouns, cannot be neatly divided into two classes, such as count versus mass. Concrete mass nouns fall into at least six groupings on formal grounds, once a range of syntactic and phraseological diagnostics are brought to bear. Second, semantic-conceptual schemas can be devised for each of the formally identifiable groupings, and the formal properties of each grouping can be aligned with particular semantic-conceptual components. Third, this approach produces not only better descriptive adequacy (a much "tighter fit" on the data), but also much greater explanatory adequacy.

These results are achievable only with an appropriate method of semantic analysis, such as that furnished by the NSM system. In particular, it has emerged that capturing the necessary level of semantic granularity requires a rather rich semantic vocabulary. It certainly cannot be done in terms of a small number of binary features. The NSM system meets this need but— importantly—without enriching the semantic metalexicon any further than is already required for semantic analysis in other lexical semantic domains. Semantic primes such as SOMETHING, PART, KIND, THIS, THE SAME, ONE, MUCH/MANY, and ALL are needed right across the lexicon. No special additions are needed to apply the existing, independently motivated metalanguage to the arena of mass expressions.

Appendix: A Semantic Account of the Conversion of Count Nouns to Mass Nouns

Many count nouns can be used also as mass nouns, and vice versa. How and why this is so has been the subject of a great deal of discussion among linguists and philosophers and has led some to question the very existence of the count/mass distinction. In this appendix I address the conversion of count nouns to mass nouns. Like Wierzbicka (1988: 522), my starting point is Apresjan's (1974) observation that rules of regular polysemy are akin to derivational rules, in that they derive new senses from old ones following regular semantic patterns. Though they do not use the term "regular polysemy," Payne and Huddleston (2002: 336–338) advance this explanation for the existence of "paired count and noncount senses" in English. They give a series of examples where one sense of a word is primarily either count or noncount and the other, secondary sense is regular and predictable by a rule of semantic extension. For example, count nouns whose primary sense is the name of a species of fish or poultry have secondary noncount senses denoting "food substances," for example, 'We're having salmon for dinner'; mass nouns denoting kinds of drinks have secondary count noun uses denoting a "serving of the drink," for example, 'She offered me another beer.'

It is crucial to recognize that such patterns of semantic extension or conversion are language specific in nature. To see this, it is enough to note that English foodstuff examples, such as 'I can't eat tomato' or 'He is allergic to mushroom,' have no mass noun counterparts in either Polish or Russian, where comparable sentences exist only with the count noun morphology (Anna Wierzbicka, personal communication). Even in English, there is no single and uniform "count to mass" pattern, but rather an array of semantically specific patterns, the most productive of which involve foodstuffs and materials. In this appendix I first discuss several of these normal or entrenched patterns and then turn to cases in which count nouns of other kinds can be coerced into mass readings.

Wierzbicka (1988: 522) observed: "[T]here is a general semantic rule in English (and in many other languages) which allows names of edible objects which have edibility encoded in their meaning to be used also as names of non-discrete foodstuffs derived from these objects." Wierzbicka is referring here to *egg, apple, cabbage, onion,* and the like, which are derived from discrete kinds of edible things by beating, chopping, grinding them, and so on. The relevant rule of regular polysemy can be stated as in schema [Q]. The initial component establishes the categorical status of the referent: 'something of one kind that people can eat [M].'[18] Component (b) is the same as found in homogeneous substance words such as *cheese, butter,* and *iron*; that is, 'people can't think about something of this kind: this thing has many parts, someone can know how many.' In the case of *egg, apple,* or *tomato* in their mass noun senses, we understand that the present state of the referent is not its natural state but is due to it having undergone some processing at an earlier time, at which time 'it was part of something else of one kind' (see components (c) and (d)). The final component, (e) then gives the word *egg (apple, capsicum, tomato,* etc.) as an identifying label for the kind of thing in question. Note that this wording manages to maintain the necessary number neutrality

---

18. The relative clause structure used in component (a) is not part of the universal syntax of semantic primes. The component can be regarded as an abbreviation for 'something of one kind; people can eat [M] things of many kinds, this is one of these kinds.'

in relation to the source items; for example, one can get some of the foodstuff *egg* from a single egg or from several eggs.

[Q] egg₂, apple₂, tomato₂:
a. something of one kind that people can eat [M]
b. people can't think about something of this kind like this:
   "this thing has many parts, someone can know how many"
c. before, it wasn't like this
d. it is like this now because someone did some things to it before
   at this time it was part of something else of one kind
e. people can say what kind with the word *egg* (*apple, tomato,* etc.)

The examples so far have chiefly concerned foodstuffs derived from plants. What about foodstuffs derived from other kinds of living things, such as animals, birds, and fish? One might expect schema [Q] would work for these also, with the prediction that the various foodstuffs would be named according to the name of the source species. This expectation is, of course, confirmed in the case of flesh deriving from birds (*chicken, turkey, quail*) and fish (*trout, bream, salmon*), but the story is more complicated for different kinds of "meat." It is a well-known peculiarity of English (one of the legacies of the Norman period) that there are specialized words—*pork, beef, mutton* (*lamb*), *venison*—for the meats derived from the traditional livestock and game, that is, pigs, cattle, sheep, and deer. On the other hand, there are no special words for meats derived from other animals (e.g., *kangaroo, moose, dog, horse*). To treat this domain from a synchronic semantic point of view, it is necessary to make use of two semantic molecules: 'meat [M]' and 'animals [M].' They are of course related, since *meat* derives from the bodies of animals.[19]

Consider schema [R]. Component (a) reads 'meat [M] of one kind.' Component (b) provides for a special word (*pork, beef,* etc.) identifying the kind of meat in question. There is no need for the customary mass or substance components, since the semantic molecule 'meat [M]' will already incorporate these components. The fact that any specific kind of meat derives from a specific kind of animal must be stated, however, and this is done in component (c). The final component then gives the identifying word for the kind of animal involved.

[R] pork (beef, dog₂, horse₂):
a. meat [M] of one kind
b. people can say what kind with the word *pork* (*beef, dog, horse*)
c. when people want there to be meat [M] of this kind somewhere,
   they do some things to animals [M] of one kind
d. people can say what kind with the word *pig* (*cow, dog, horse*)

---

19. The concept of *meat* presumably includes such components as 'some soft [M] parts of the bodies of animals [M]; people can eat [M] these parts, if they do some things to them beforehand.'

If this is correct, the semantic structure of "meat words" is different from that of other foodstuffs.[20]

A third suite of productive English schemas concerns "materials," that is, kinds of things people use to make things of other kinds, which are derived from natural sources. One such schema, which relates the names of trees to the names of timbers obtained from these trees, is given in [S]. The schema employs the semantic molecule 'wood [M].'[21] It is a bit simpler than the meat schema, because the different kinds of wood (*cedar, pine, oak*, etc.) invariably go by the same name as the kind of tree from which they are obtained.

[S] **cedar$_2$ (pine$_2$, oak$_2$, etc.):**
a. wood [M] of one kind
b. when people want there to be wood [M] of this kind somewhere,
    they do some things to trees [M] of one kind
c. people can say what kind with the word *cedar* (*pine, oak*)

Can this schema be generalized to apply to other kinds of plant products derived from plants, such as *cotton, hemp, tobacco,* and *flax*? In my view such a move would not be justified, if only because for these words the product term is intuitively the primary one, while the plant name is the derived or secondary term; that is, the direction of the lexical extension is the opposite to that with kinds of trees and the kinds of wood deriving from them. Also, there is no superordinate (comparable to 'wood [M]') for materials in general (the word *material* itself is a functional category, comparable to *foodstuff*).

In general, linguists have been less interested in the regular and routine patterns just illustrated than in creative and coerced usages, such as Pelletier's (1979: 7) famous sentence about the mother termite who says about her child: "Johnny is very choosey about his food. He will eat book, but he won't touch shelf." Such atypical examples are important because they confirm that countability cannot be regarded as a fixed lexical or syntactic property but is, on the contrary, "highly sensitive to the intended conceptualization" (Wierzbicka 1988: 507). In my view, however, the relative disinterest in the patterns of regular polysemy has meant overlooking important aspects of the mechanism involved with the atypical examples. It is no coincidence that the termite sentence is about foodstuffs: the words *book*

---

20. Other foodstuffs do not fall under superordinate categories; that is, though *meat* is a genuine superordinate of *pork, beef*, and so on, *vegetable* is not a genuine superordinate of *potato, cabbage*, and so forth, nor is *fruit* a genuine superordinate of, for example, *apple, orange*. This is because *vegetables, fruit, cereals*, and such are collective terms, not sortals (Wierzbicka 1985, 1988).

21. The concept of *wood*$_1$ presumably includes such components as 'people make[M] things of many kinds from things of this kind, the hard [M] parts of a tree [M] are something of this kind.' The word *wood* is polysemous, with the other meaning *wood*$_2$ relating to fire. There are lots of compounds that select one or other of these meanings, for example, *woodwork* and related words such as *timber*, on the one hand, and *firewood, woodpile*, and *driftwood*, on the other.

and *shelf* are assimilating to mass noun status via the foodstuff schema given in [Q]. Putting it another way, we understand the novel intended meanings of *book* and *shelf* by analogy with the entrenched foodstuff pattern.[22]

What, then, are we to make of so-called "grinder" examples like Pelletier's (1979) sentence (a), or Jackendoff's (1992) sentence (b) below? Pelletier used the (a) sentence to illustrate his conception of an imagined "universal grinder" which would reduce any object fed into it to "the finely-ground matter of which it is composed." Jackendoff's example could apply to the situation after a nasty car accident in the suburbs. Payne and Huddleston (2002) use a similar example, given in (b'). Notably, all three examples are of the "distributed location" variety.

a. So a hat is entered into the grinder and after a few minutes there is hat all over the floor.
b. There was dog all over the road.
b'. There was cat all over the driveway.

Bloomfield's (1933: 205) sentence 'He got egg on his necktie' arguably manifests the same pattern. This sentence might first be thought to be covered by the foodstuff schema in [Q], since one can readily imagine someone who has been careless over breakfast ending up in the situation described. But someone who says such a sentence is not thinking of the *egg* as a foodstuff, but rather as a substance (compare 'He had blood on his tie'). The sentence 'He had egg on his tie' is equally applicable to a situation in which someone has been pelted by eggs; furthermore, in both contexts it is possible to imagine the extension of the word *egg* including some bits of egg shell, which would not be permissible for *egg* in the foodstuff sense. A similar point applies to Jackendoff's sentence about the aftermath of a road accident ('There was dog all over the road') compared with the use of the word *dog* as a meat word, for example, 'We had no idea that it was dog' (said, e.g., after eating in a restaurant in Vietnam). After an accident, the stuff referred to as *dog* presumably includes bones and fur, which would not be possible in the meat sense.

In summary, the mechanisms of semantic coercion depend on the existence of established mass noun schemas. Hearing a word like *book*, *dog*, or *hat* used without the expected count noun morphology cues the listener to search for an interpretation that makes sense. The range of available interpretations depends on established semantic schemas and rules of regular lexical polysemy and is therefore partly language specific.

---

22. The relationship between formal properties and conceptual components is "bidirectional" (Middleton et al. 2004). I therefore agree with Bloom (2000: 198–211) and others who have argued that young children can use the syntactic behavior of mass nouns as a guide to their meanings. Just as meaning is a guide to form, so, too, form is a guide to meaning.

ACKNOWLEDGEMENTS

The explications were codeveloped with Anna Wierzbicka. I am grateful to Jock Wong, Brett Baker, and Mee Wun Lee for useful comments and input. This chapter has benefited from input received from participants in the New Directions in Cognitive Science conference held in February 2006 at Simon Fraser University, Vancouver. I am also grateful for helpful comments from two anonymous reviewers. This research was supported by the Australian Research Council.

REFERENCES

Apresjan, Juri D. 1974. Regular polysemy. *Linguistics* 142, 5–32.
Bloom, Paul. 2000. *How Children Learn the Meaning of Words*. Cambridge, MA: MIT Press.
Bloomfield, Leonard. 1933. *Language*. London: Allen and Unwin.
Climent, Salvador. 1996. Semantics of portions and partitive nouns for NLP. In *Proceedings of the 16th Conference on Computational Linguistics (Copenhagen, Denmark)* (pp. 243–248). Morristown, NJ: Association for Computational Linguistics.
Cruse, D.A. 1986. *Lexical Semantics*. Cambridge, UK: Cambridge University Press.
Gillon, Brendan S. 1992. Towards a common semantics for English count and mass nouns. *Linguistics and Philosophy* 15, 597–639.
Goddard, Cliff. In press. The conceptual semantics of numbers and counting: An NSM analysis. *Functions of Language* 16(2).
———. 2007a. Semantic primes and conceptual ontology. In Andrea Schalley and Dietmar Zaefferer (eds.), *Ontolinguistics: How Ontological Status Shapes the Linguistic Coding of Concepts* (pp. 145–174). Berlin: Mouton de Gruyter.
———. 2007b. Semantic molecules. In Ilana Mushin and Mary Laughren (eds.), *Selected Papers of the 2006 Annual Meeting of the Australian Linguistic Society*. Available at http://www.als.asn.au.
———. 2006a. Natural semantic metalanguage. In Keith Brown (ed.), *Encyclopedia of Language and Linguistics*, 2nd ed. (pp. 544–551). Oxford: Elsevier.
———. 2006b. "Lift your game, Martina!"—deadpan jocular irony and the ethnopragmatics of Australian English. In Cliff Goddard (ed.), *Ethnopragmatics: Understanding Discourse in Cultural Context* (pp. 65–97). Berlin: Mouton de Gruyter.
———. 1998. *Semantic Analysis*. Oxford: Oxford University Press.
Goddard, Cliff, and Wierzbicka, Anna. 2007. NSM analyses of the semantics of physical qualities: *Sweet, hot, hard, heavy, rough, sharp* in cross-linguistic perspective. *Studies in Language* 31(4), 765–800.
Goddard, Cliff, and Wierzbicka, Anna (eds.). 2002. *Meaning and Universal Grammar—Theory and Empirical Findings*. Vols. 1 and 2. Amsterdam: John Benjamins.
Jackendoff, Ray. 2002. *Foundations of Language*. Oxford: Oxford University Press.

———. 1992. Parts and boundaries. In Beth Levin and Steven Pinker (eds.) *Lexical and Conceptual Semantics* (pp. 9–46). Oxford: Blackwell.
———. 1990. *Semantic Structures*. Cambridge, MA: MIT Press.
Levin, Beth, and Rappaport Hovav, Malka. 2005. *Argument Realization*. Cambridge: Cambridge University Press.
Middleton, Erica L., Wisniewski, Edward J., Trindel, Kelly A., and Imai, Mutsumi. 2004. Separating the chaff from the oats: Evidence for a conceptual distinction between count noun and mass noun aggregates. *Journal of Memory and Language* 50, 371–394.
Payne, John, and Huddleston, Rodney. 2002. Nouns and noun phrases. In Rodney Huddleston and Geoffrey Pullum (eds.), *The Cambridge Grammar of the English Language* (pp. 323–523). Cambridge: Cambridge University Press.
Pelletier, F.J. 1979. Non-singular reference: Some preliminaries. In F.J. Pelletier (ed.), *Mass Terms: Some Philosophical Problems* (pp. 1–14). Dordrecht: Reidel. (Originally published in 1975 in *Philosophia* S(4): 451–465.)
Pelletier, F.J., and Schubert, L. 2003. Mass expressions. In D. Gabbay and F. Guenthner (eds.), *Handbook of Philosophical Logic* (2nd ed., Vol. 10, pp. 249–335). Dordrecht: Kluwer Academic Publishers.
Pustejovsky, James. 1995. *The Generative Lexicon*. Cambridge, MA: MIT Press.
Tien, Adrian. 2005. The semantics of children's Mandarin Chinese: The first four years. PhD thesis, University of New England, Armidale, Australia.
Wierzbicka, Anna. In press. The theory of the mental lexicon. In Sebastian Kempgen, Peter Kosta, Tilman Berger, and Karl Gutschmidt (eds.), *Die Slavischen Sprachen/The Slavic Languages: Ein internationales Handbuch zu ihrer Struktur, ihrer Geschichte und ihrer Erforschung/An International Handbook of their Structure, their History and their Investigation*. (Handbücher zur Sprach- und Kommunikationswissenschft/Handbooks of Linguistics and Communication Science HSK 32(1)). Berlin: Mouton de Gruyter.
———. 2006a. Shape in grammar revisited. *Studies in Language* 30(1), 115–177.
———. 2006b. The semantics of colour: A new paradigm. In Carole P. Biggam and Christian J. Kay (eds.), *Progress in Colour Studies. Volume 1: Language and Culture* (pp. 1–24). Amsterdam: John Benjamins.
———. 1996. *Semantics: Primes and Universals*. Oxford: Oxford University Press.
———. 1991. Semantic rules know no exceptions. *Studies in Language* 15(2), 371–398.
———. 1988. Oats and wheat: Mass nouns, iconicity, and human categorisation. In *The Semantics of Grammar* (pp. 499–560). Amsterdam: John Benjamins.
———. 1985. *Lexicography and Conceptual Analysis*. Ann Arbour: Karoma.
———. 1980. *Lingua Mentalis: The Semantics of Natural Language*. Sydney: Academic Press.
———. 1972. *Semantic Primitives*. Translated by Anna Wierzbicka and John Besemeres. Frankfurt: Athenäum Verlag.
Wisniewski, Edward J., Lamb, Christopher A., and Middleton, Erica L. 2003. On the conceptual basis for the count and mass noun distinction. *Language and Cognitive Processes* 18(5/6), 583–624.

Wisniewski, Edward J., Imai, Mutsumi, and Casey, Lyman. 1996. On the equivalence of superordinate concepts. *Cognition* 60, 269–298.

Xu, Fei. 2005. Categories, kinds and object individuation in infancy. In Lisa Gershkoff-Stowe and David H. Rakison (eds.), *Building Object Categories in Developmental Time* (pp. 63–89). Mahwah, NJ: Lawrence Erlbaum.

9

On Using Count Nouns, Mass Nouns, and Pluralia Tantum: *What Counts?*

EDWARD J. WISNIEWSKI

English and other languages make a grammatical distinction between count nouns and mass nouns. For example, "dog" is primarily used as a count noun, and "mud" is primarily used as a mass noun. Count nouns but not mass nouns can be pluralized and preceded by numerals (as in "three dogs" but not "three muds"). Count nouns but not mass nouns can appear with the indefinite determiner "a" (as in "A dog ate the chicken" but not "A mud covered the chicken"). On the other hand, mass nouns can appear with indefinite quantifiers such as "much" or "little" (as in "much mud" but not "much dog"), whereas count nouns can appear with indefinite quantifiers such as "many" and "few" (as in "many dogs" but not "many muds").[1]

Many scholars have addressed the question of why there is a linguistic distinction between count nouns and mass nouns. Put another way, why do speakers use count and mass nouns in a language? If one broadly examines the use of count and mass nouns in English, the answer to this

---

1. Some languages such as classifier languages (e.g., Japanese) do not make a distinction between count and mass nouns. Nevertheless, they do have other mechanisms for indicating that an entity is or is not individuated (for discussions of cross-linguistic differences in the tendency to individuate, see Imai and Gentner, 1997; Lucy, 1992; Wisniewski et al., 2003).

question is not straightforward. Count nouns and mass nouns refer to diverse things, such as abstract entities (e.g., an idea vs. insanity), events (an explosion vs. sleep), and sounds (e.g., a knock vs. thunder). Count and mass nouns also refer to perceptually similar things (e.g., pebbles vs. gravel, leaves vs. foliage, garments vs. clothing, and advice vs. suggestions). In addition, many nouns are readily used either as a count or mass noun (e.g., "I'll go buy a cake" vs. "I'll have some cake," "She is a curiosity" vs. "She is full of curiosity"). These observations reveal a paradox: the count—mass noun distinction is made across diverse domains that appear to have little in common and within domains that appear to have much in common.

Is there a conceptual basis to the grammatical distinction between count nouns and mass nouns? One answer is that this grammatical distinction is, to a very large degree, semantically opaque and unprincipled (e.g., Bloomfield, 1933; Gleason, 1969; Markman, 1985; McCawley, 1975; Palmer, 1971; Quine, 1960; Ware, 1979; Whorf, 1962). In general, people learn which nouns are typically used as count nouns and which are typically used as mass nouns without any understanding of *why* these differences in syntax occur. Another answer is that the grammatical distinction between count and mass nouns is to a very large degree conceptually based. That is, when speakers use count nouns to refer to things they implicitly have something in mind that they are trying to communicate that is common across all uses of count nouns. A similar view applies to the use of mass nouns. A third answer, and the one that I propose, is that the count—mass noun distinction is to a very large degree conceptually based, but there are exceptions. Some exceptions do not seem to have a clear explanation, but others may occur because of *competing communicative functions* of language.

The remainder of this chapter is organized in the following manner. I first describe the *cognitive individuation hypothesis*, which proposes that there is a conceptual basis for the count—mass noun distinction. Next I describe empirical evidence that is consistent with this hypothesis. Specifically, I describe experiments and linguistic analyses that evaluate the conceptual basis for the count—mass noun distinction across a diverse set of domains: aggregates (groups of small, relatively homogeneous entities, e.g., rice and toothpicks), superordinates (broad categories of perceptually heterogeneous entities, e.g., vehicles and clothing), and sounds. I also examine the conceptual basis for the use of collective nouns (singular nouns, e.g., *team and constellation*, that refer to multiple entities) and *pluralia tantum* (plural nouns, e.g., *groceries* and *ruins*, that refer to multiple entities or entity-like things). I then describe uses of count—mass noun syntax that contradict the cognitive individuation hypothesis. Some of these exceptions evidently arise because of the competing communicative functions of language. Finally, I describe the broader relationship between the count—mass noun distinction and cognition and perception.

The Cognitive Individuation Hypothesis

A number of scholars have proposed the cognitive individuation hypothesis as an explanation for a conceptual basis of the count—mass noun distinction (Bloom, 1990, 1996; Langacker, 1987; Mufwene, 1984; Soja et al., 1991; Wierzbicka, 1988; Wisniewski et al., 1996). According to this hypothesis, count nouns refer to individuals, and mass nouns refer to nonindividuated entities. For example, objects are prototypical individuals in being discrete, bounded entities that are separate from other aspects of the world. Substances are prototypical nonindividuated entities in being continuous, unbounded, and arbitrarily divisible (e.g., mud divided into a portion of any size is still mud). In corresponding fashion, most objects are labeled with count nouns (e.g., *a cat, a computer*), and most substances with mass nouns (e.g., *clay, honey*).

There are two other important aspects of the cognitive individuation view. First, the idea of an individual is broader than an object and includes more abstract individuals. For example, *constellation* refers to an abstract individual. Even though it is composed of objects (i.e., stars), the cognitive individuation view claims that people think of these stars as a single individual. Likewise, the idea of a nonindividuated entity is broader than substances and includes more abstract nonindividuated entities, such as curiosity, which can refer to an abstract quantity of a varying amount.

Second, the cognitive individuation view emphasizes the role of the human or cognitive agent in determining whether perceptual input from the world is interpreted or construed as an individual or as a nonindividuated entity. How we interact with things, our goals, and our focus of attention influence our interpretation of the perceptual input and affect whether we refer to something with a count noun or a mass noun. For example, in most contexts people interpret a bench as a distinct individual and hence refer to it as "a bench." The perceptual input readily gives rise to the perception of a discrete, bounded object. However, a person who wants to sit on a crowded bench might tell those sitting there to "move over and give me some bench." In this context, the person conceptualizes the bench as a nonindividuated entity (i.e., a flat expanse or quantity of space) and hence uses a mass noun.

The cognitive individuation view straightforwardly implies that perceptually similar entities may be conceptualized as individuated or as nonindividuated depending on the cognitive agent. This observation is important because scholars who suggest that the count—mass noun distinction is unprincipled cite pairs of count and mass nouns whose referents are perceptually similar (oats vs. wheat, Bloomfield, 1933; rice vs. beans, Gleason, 1969; pebbles vs. gravel, Markman, 1985; noodles vs. spaghetti, McCawley, 1975; foliage vs. leaves, Palmer, 1971; footwear vs. shoes, fuzz vs. cop, fruit vs. vegetable, Ware, 1979). However, these scholars have focused on the perceptual similarity of these entities without taking into consideration the role of the cognitive agent.

The cognitive individuation view raises the question of what is meant by an individuated or nonindividuated entity. Wisniewski et al. (2003) suggest that the distinction between an individual and a nonindividuated entity is determined by its *scope of predication*. People conceptualize an entity as an individual if they are able to predicate its central properties to that entity "as a whole" rather than to arbitrary portions or parts of that entity. For example, we typically conceptualize "postage stamp" as individuated. In turn, the central properties of a postage stamp, such as "is adhesive, is torn from a sheet, is used for mailing a letter, and is rectangular," do not apply to arbitrary portions of a stamp. In contrast, people will consider an entity as nonindividuated if they are able to predicate it central properties to the entity that apply to arbitrary portions or parts of that entity. For example, we typically conceptualize "butter" as nonindividuated. In turn, central properties of butter, such as "melts, has a particular taste, used as a topping for food," largely apply to arbitrary portions of butter. This hypothesis raises the question of what is meant by central properties. Roughly speaking, they are properties that are responsible for an entity "being what it is," such that it would be difficult to imagine that entity existing without such properties" (for a candidate operational definition of this idea, see Sloman et al., 1998).

Evidence for the Cognitive Individuation Hypothesis

A number of psychological studies and linguistic analyses are consistent with the cognitive individuation hypothesis. In this section, I describe evidence from cognitive psychology and linguistics suggesting that speakers conceptually distinguish between the referents of count and mass nouns. The evidence spans a number of different kinds of categories (aggregates, superordinates, and sounds) and types of linguistic terms (collective nouns and *pluralia tantum*).

Aggregates

These entities consist of multiple, co-occurring, relatively small, homogeneous constituents. For example, rice is an aggregate. The term *rice* typically refers to a group of grains of rice, stored uncooked in bags and other containers, or appearing in cooked form in pots or bowls, or on plates. There are both aggregates named by count nouns (e.g., *grapes, toothpicks, tacks*) and aggregates named by mass nouns (e.g., *sugar, grass, sand*). Some pairs of aggregates consist of very similar elements even though one member of a pair is named by a count noun and the other by a mass noun (e.g., *pebbles and gravel, snow flurries* and *snow, pills* and *aspirin, coins* and *change*). Thus, it is not readily apparent whether a conceptual distinction exists between count- and mass-noun aggregates.

Wierzbicka (1988) proposed several reasons for why a conceptual distinction exists between count- and mass-noun aggregates. She suggested that differences in how speakers interact with the constituents of an aggregate determine whether they will individuate those constituents, which in turn will be reflected in the syntax of the language. For example, Wierzbicka notes that in Polish, groups of berrylike fruits (e.g., raspberries, currants, strawberries, and plums) are named by plural count nouns perhaps because Polish people usually interact with the constituents one by one (e.g., when picking them or eating them). In contrast, Polish farmers selling their wares at a market setting commonly refer to such fruits with mass nouns perhaps because they interact with them as quantities and not as individual entities. Wierzbicka also suggested that the ease of distinguishing the constituents of an aggregate influences whether a speaker construes those constituents as individuals. For example, she argues that "beans" is a plural count noun and "rice" a mass noun because beans are more perceptually distinguishable than are individual grains of rice. Thus, people view each bean as a distinct individual but rice as a nonindividuated group.

In a number of studies, Middleton et al. (2004) evaluated these hypotheses proposed by Wierzbicka. One experiment assessed whether perceptual distinguishability of elements composing familiar aggregates predicted their count—mass noun status. The experiment involved a large, diverse set of 112 aggregates. Undergraduates simply rated how easy it is to see or distinguish the individual elements that compose each aggregate, using a numerical scale ranging from 1 to 9, with 1 indicating that the elements were extremely easy to see and distinguish and 9 indicating that the elements were extremely hard to see and distinguish. (Numbers in between these extremes indicated intermediate degrees of perceptual distinguishability.) Each count-noun aggregate was presented in the format "a(n) X" (e.g., a bean), and every mass-noun aggregate was presented in the format "a unit of X" (e.g., a unit of rice). The referents of these phrases are single elements of an aggregate. For each element, participants were instructed to visualize a typical group of that element before making their rating. This procedure minimizes the effects of count—mass noun syntax on participants' judgments. For example, participants' judgments about the size of the elements of "rice" or "beans" could be influenced by their knowledge that rice is a mass noun and beans is a count noun.

Consistent with Wierzbicka's hypothesis, the rated ease of distinguishing the elements of an aggregate was systematically related to whether the aggregate was named by a count or mass noun. Undergraduates gave lower ratings to count-noun aggregates than to mass-noun aggregates. (There were some exceptions to this result that I discuss further below.)

In another study in Middleton et al. (2004), undergraduates rated how often they interacted with one or a few of the elements of an aggregate, using a numerical scale ranging from 1 to 9, with 1 indicating that they frequently interacted with one or a few of the individual elements and

9 indicating that they rarely interacted with one or a few of the individual elements of the aggregate. (Numbers between these extremes indicated intermediate degrees of interaction with one or a few of the individual elements.) Consistent with Wierzbicka's hypothesis, the rated likelihood of interacting with one or a few elements of an aggregate was systematically related to whether the aggregate was named by a count or mass noun: undergraduates gave lower ratings to count-noun aggregates than to mass-noun aggregates. (Again, there were some exceptions to this result that I discuss further below.)

To assess the generality of these findings, Middleton et al. (2004) conducted two additional studies that also involved ratings of perceptual distinguishability of aggregate elements and likelihood of interacting with one or a few versus multiple elements, and used another set of 80 aggregates that did not overlap with the first set and replicated the previous results.

These findings show that people are aware of differences between the referents of familiar count and mass-noun aggregates. That is, compared to the elements of mass-noun aggregates, people believe that the elements comprising familiar count-noun aggregates are more perceptually distinguishable and that their interactions with the elements more often involve one or a few elements at a time. However, these studies do not show that people's knowledge of these differences *causes* them to conceptualize an aggregate as either individuated or as nonindividuated and consequently to name the aggregate with either a count noun or mass noun. To obtain more direct evidence for differences in conceptualization and their consequences, Middleton et al. (2004) conducted two other experiments in which undergraduates determined which of two novel aggregates was more likely to be referred to by a novel count noun (or in other cases, a novel mass noun).

In one experiment, the perceptual distinguishability of the elements making up novel aggregates varied. Undergraduates saw a series of pairs of novel aggregates accompanied by a phrase containing either a novel count or mass noun. (Pictures of novel aggregates were actually used.) Each pair of aggregates differed either in the size of their elements (relatively small versus large), in the spatial proximity of their elements (relatively close together or far apart), or along both dimensions. Figure 9.1 presents examples of each pair along with a novel count or mass-noun phrase that was presented with the pair. (Some participants saw a pair with a count-noun phrase, but other participants saw the same pair with a mass-noun phrase.) We assumed that the greater size of elements and the greater distance between elements increase their distinguishability. A participant's task was to read the phrase below the pair of aggregates and to assume that someone had said the phrase when talking about one of the two aggregates. The participant was to pick the aggregate that this someone was likely to be talking about.

The results showed a strong effect of spatial contiguity. Consistent with our hypothesis, participants frequently chose novel count nouns as names

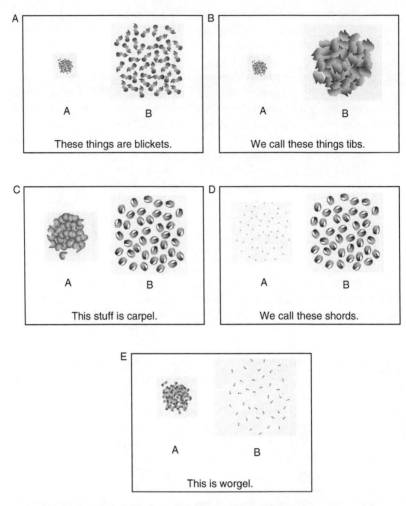

FIGURE 9.1 Examples of pairs of novel aggregates and phrases referring either to a novel count noun or to a novel mass noun (Middleton et al., 2004)

for aggregates with noncontiguous elements but chose novel mass nouns as names for aggregates with contiguous elements. For example, figure 9.1E shows an aggregate with small elements close together (small-close) on the left and an aggregate with small elements relatively far apart (small-apart) on the right. Participants tended to choose the aggregate on the right for phrases containing novel count nouns (e.g., 'These things are blickets'). However, other participants tended to choose the aggregate on the left for phrases containing novel mass nouns (e.g., 'This stuff is blicket'). On the other hand, size did not matter. That is, participants did not tend to pick aggregates with large elements over those with small elements when a phrase contained a novel count noun (and vice versa, when a phrase contained a novel mass noun). In hindsight, we noticed that the elements of the small-noncontiguous aggregates are physically separate and can be seen as distinct from each other, even though they are relatively small. This factor may explain the lack of a size effect for the small-apart versus large-apart comparison (figure 9.1D). Conversely, for the small-together versus large-together comparison, it appears difficult to perceptually separate the elements in either aggregate, whether they are large or small (figure 9.1B). This factor may explain the lack of size effect for this comparison. These findings provide direct evidence that perceptual distinguishability affects how people conceptualize aggregates.

In a final experiment, Middleton et al. (2004) examined whether interacting with a novel aggregate would *change* the way that undergraduates *conceptualize* the aggregate. In a baseline condition (not involving interaction), participants examined a novel aggregate consisting of yellow, coarse-grain sugar formed into a circular-shaped pile inside an open box (see figure 9.2, top). They were asked to choose which of two novel phrases someone might say when referring to the contents of the box. One phrase contained a novel plural count noun (e.g., 'We call these blickets') and another contained a novel mass noun (e.g., 'We call this blicket'). A majority of the participants (61%) selected a mass-noun phrase as best describing the aggregate.

In the interaction condition, other participants saw the same yellow coarse-grain sugar presented in the same box as in the baseline condition. However, they also interacted with single elements of the aggregates. In particular, participants were each given a board with holes big enough for a grain of sand to be dropped through and a small metal implement that enabled them to pick up a single grain of sugar and to drop it through one of the holes (see figure 9.2, bottom). The experimenter illustrated the task by carrying it out herself for two minutes and then signaling the subjects to begin with their own boards. The participants then carried out the task for fifteen minutes. Then, as in the baseline condition, participants were asked to choose which of two novel phrases someone might say when referring to the contents of the box. In contrast to the baseline condition, a majority (69%) of the participants selected a novel plural count-noun phrase as best describing the aggregate even though they observed the same identically displayed aggregate as did subjects in the baseline condition.

This study provides direct evidence that how people interact with an aggregate affects their conceptualization of that aggregate as individuated or as nonindividuated. When participants simply observed an aggregate consisting of many very small, spatially contiguous elements, they construed the aggregate as a nonindividuated group. In turn, given a choice between labeling the aggregate with a count or a mass noun, participants chose a mass noun (just as they did with this type of aggregate in the previous study). However, interacting with the elements of this aggregate

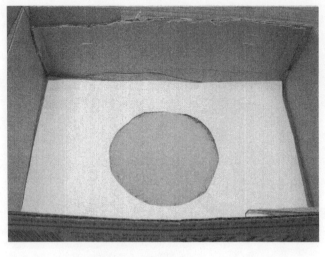

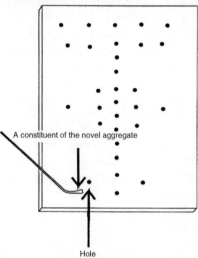

FIGURE 9.2 Novel aggregate presented to participants (top) and board and implement for the interaction task (bottom)

one at a time led participants to conceptualize the aggregate as a collection of distinct individuals. In turn, given a choice between labeling the aggregate with a count or a mass noun, they chose a count noun. Thus, participants conceptualized a perceptually identical entity in different ways depending on whether they interacted with that entity. These findings are consistent with the cognitive individuation hypothesis.

Taken together, the findings from these six experiments suggest that important properties of count-noun aggregates individually apply to each aggregate element, whereas those of mass-noun aggregates apply to arbitrary-sized groups of elements. Thus, the two types of aggregates differ in their scope of predication. For example, the interaction "cleaning teeth" applies individually to each toothpick at a time and not to the aggregate of toothpicks at the same time, and "buttoning" applies individually to each button at a time and not to the aggregate of buttons at the same time. In contrast, for example, 'flavors food' applies to multiple grains of pepper or salt at a time and not to a single grain. Evidently, one can (to a large degree) predicate these properties of arbitrary size portions of a mass-noun aggregate (though we did not directly test this part of the hypothesis). For instance, there are a variety of amounts of pepper or salt that one could use to flavor food.

Superordinate Categories

Superordinates refer to broad categories of perceptually diverse things. For example, *vehicle* is a superordinate category whose members include car, bicycle, truck, balloon, boat, airplane, train, helicopter, and motorcycle. There are both count-noun superordinates (e.g., VEHICLES, ANIMALS, PLANTS, and TOOLS) and mass-noun superordinates (e.g., FURNITURE, CLOTHING, VEGETATION, and SKI GEAR). Both count-noun and mass-noun superordinates appear to refer to discrete objects—prototypical, individuated entities. For example, the referents of *shirt*, *coat*, and *sweater* are typically conceptualized as distinct individuals. Yet, they are also considered CLOTHING (a mass-noun superordinate). Why is there a grammatical distinction between count- and mass-noun superordinates if the members of all superordinates appear to be distinct individuals? Largely because of this paradox, researchers have suggested that the distinction between count- and mass-noun superordinates is not conceptually based (Gordon, 1985; Markman, 1985; McPherson, 1991; Murphy and Wisniewski, 1989; Ware, 1979).

In addressing this paradox, Wisniewski et al. (1996) suggested that people conceptualize count-noun and mass-noun superordinates differently. Specifically, people use count-noun superordinates to refer to one or more distinct individuals, each of which is a member of the count-noun superordinate category. In contrast, people use a mass-noun superordinate to refer to a nonindividuated group of multiple entities. (These differences in conceptualization are somewhat analogous to those that characterize count- and mass-noun aggregates.) For example, consider the count-noun superordinate VEHICLE and the mass-noun superordinate FURNITURE. When someone says, 'The new truck made by Ford is a gas guzzling vehicle,' that person is conceptualizing the new truck as an individual that is a member of the category of vehicles. When someone says, 'Tow those vehicles!' when referring to a truck and a car that are illegally parked, that person considers each one of those to be a member of the category VEHICLE. Hence, the referent of vehicles is an "individuated group." But, when someone using the mass noun FURNITURE says, 'The furniture makes the living room look cluttered,' in the presence of two couches, several lamps, and a table and chairs, that person is not conceptualizing each of these entities as a member of the category of furniture. That is, in this context, the person is not thinking of an individual couch as furniture or an individual chair as furniture, and so forth. Rather the person is thinking of the couches, lamps, table, and chairs *together* as furniture.

As a consequence of this difference in conceptualization, a count-noun superordinate is a true taxonomic category, whereas a mass-noun superordinate is not a true taxonomic category (see also Bloom, 1990; Wierzbicka, 1988). Entities associated with a taxonomic count-noun category are each a member of the category and can inherit properties of the category that apply to each individual member. For example, if you are told that an opprobine is a vehicle, then you can infer that a particular opprobine probably has a steering wheel, is operated by one or two people, goes from one location to another, and so on. Figure 9.3 illustrates this hypothesized difference in conceptualization.

Wisniewski et al. (1996) conducted a number of studies in an attempt to provide converging evidence for this view, using a relatively large number of superordinates (twenty to forty per experiment). In a property inference task, undergraduates listed those properties that tended to characterize a superordinate category. Many of the properties could be classified as human interactions with superordinate categories (e.g., "eat" for food, "get sick from" for disease, "you play" for musical instrument). These properties are central to most superordinates. If people conceptualize a member of a count-noun superordinate as a single entity but conceptualize a *group* of entities as

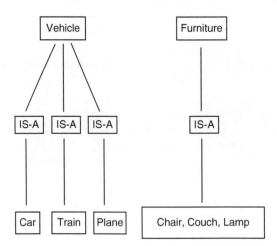

FIGURE 9.3   Illustration of conceptual differences between count- and mass-noun superordinates

a member of a mass-noun superordinate, then count- and mass-noun superordinates should differ in their scope of predication. Thus, we predicted that the interactions listed for count-noun superordinates would more often involve a single entity of a count-noun superordinate, whereas the interactions listed for mass-noun superordinates would more often involve multiple entities.

To test this prediction, a different group of undergraduates judged whether each of these interactions involved mostly one or mostly more than one entity associated with the superordinate category. There was also an in-between option—"sometimes one and sometimes more than one." The findings were consistent with our predictions. The average proportion of judgments that interactions involved mostly one entity was .43 for count-noun superordinates (e.g., "you play" for musical instrument, "get sick from" for disease) compared to .22 for mass-noun superordinates. The average proportion of judgments that interactions involved mostly more than one entity was .48 for mass-noun superordinates (e.g., "eat" for food, "wash" for clothing) compared to .25 for count-noun superordinates.

Participants in the property inference task also often listed parts of objects (e.g., "has windshield wipers" for "vehicle," "has a handle" for "tool," "has legs" for "animal"). These properties apply *separately to a single entity* rather than to a group. For example, "has windshield wipers" does not apply to a group of vehicles as a whole but to each individual vehicle. Likewise, the property "has a label" does not apply to a group of clothing as a whole but to each item of clothing. If people conceptualize a member of a count-noun superordinate as a single entity but conceptualize a *group* of entities as a member of a mass-noun superordinate, then count- and mass-noun superordinates should differ in their scope of predication with respect to these properties. Thus, Wisniewski et al. (1996) predicted that object parts should be listed more often for count-noun superordinates than for mass-noun superordinates. This prediction was confirmed—participants listed a greater number (and proportion) of object parts as properties of count-noun superordinates.

In the property inference task just described, undergraduates were given the names of count- and mass-noun superordinates and were instructed to list as many properties that tended to characterize the members of the superordinates. In a second property inference task, Wisniewski et al. (1996) presented undergraduates with entities that are typically associated with count-noun superordinates (e.g., "school" for building, "violin" for musical instrument) and with mass-noun superordinates (e.g., "tent" for camping equipment, "necklace" for jewelry), as well as entities that were atypical of the superordinates (e.g., "barn" for building, "lantern" for camping equipment). Participants were instructed to list the four things "that were most essential to knowing about the entity." If people think of an entity as individually belonging to a count-noun superordinate category but multiple entities together as belonging to mass-noun superordinate categories, then they should more often list "membership in a superordinate category" as an important property of an entity belonging to a count-noun superordinate category. For example, if people think of a single car as an individual vehicle then they should list "is a vehicle" as an important property of car. However, if people do not think of a single shirt as an individual item of clothing but rather think of shirt, pants, and socks together as clothing then they should be less likely to list "is clothing" as an important property of shirt. These predictions were confirmed. Participants listed membership in the superordinate category almost four times as frequently for a single entity of a count-noun superordinate than for a single entity of a mass-noun superordinate (38% vs. 10%, respectively).

A final task examined category verification times for a single entity versus multiple entities associated with a superordinate category. Subjects more quickly verified that a single item belonged to a count-noun superordinate than to a mass-noun superordinate but were slightly faster to judge that several items belonged to a mass-noun superordinate than to a count-noun superordinate. For example, participants were faster to verify sentences such as "A bomb is a kind of weapon" (single entity, count-noun superordinate) than ones such as "A chair is a kind of furniture" (single item, mass-noun superordinate), but they were slightly faster to verify sentences such as "Chair and table are furniture" (several entities, mass-noun superordinate) than ones such as "A gun and a bomb are weapons" (several entities, count-noun superordinate). Again, these results suggest that people conceptualize members of a count-noun superordinate as individually belonging to that category. But, they conceptualize a group of entities together as belonging to a mass-noun superordinate category.

Taken together, our findings suggest that count-noun superordinates but not mass-noun superordinates are true taxonomic categories. However, "members" of mass-noun superordinates are groups of entities, and properties of these superordinates more often characterize the group rather than applying separately to each entity of that group. Thus, it is misleading to refer to one or more entities as individually belonging to a mass-noun superordinate category. The findings also underscore the importance of the cognitive agent in determining the reference of a term. Prima facie, the referents of both count and mass-noun superordinates appear equivalent in generally being prototypical objects. This apparent equivalency is one reason for the view that the count–mass noun distinction between superordinates is not conceptually motivated. However, the results of the first property inference task suggest that people's conceptualizations of these entities are different, arising at least in part because of differences in how the cognitive agent interacts with those entities.

Sounds

People refer to sounds by using count nouns (e.g., a shriek, a click) as well as mass nouns (e.g., thunder, yelling). Why do we apparently individuate some sounds and not others? Little empirical work has addressed this issue. One exception is a study by Bloom (1990). He examined whether the use of count- and mass-noun syntax was sensitive to temporal properties of novel sounds. In particular, undergraduates read descriptions of two types of sounds that were produced by a machine. One sound was described as a series of temporal events with silence between, and the other as one long continuous sound:

> *Sound-discreet intervals*: John attached the moop-producer to his stereo and switched it on. It started off very loud and then was silent. Then loud, then silent. This went on for many hours.
> *Sound-continuous interval*: John attached the moop-producer to his stereo and switched it on. It started off very loud and stayed loud; the volume never changed. This went on for many hours.

Participants then chose a sentence that best described the sound. Bloom found that participants preferred sentences referring to novel count nouns for the discrete sounds (e.g., "The machine produced a lot of moops") but sentences referring to novel mass nouns for the continuous sounds (e.g., "The machine produced a lot of moop").

Sounds have a number of temporal properties that could be related to individuation: onset, offset, and duration of a sound, and the inter-sound interval of silence in the case of multiple occurrences of a sound. A sound may be individuated to the extent that its onset and offset are rapid and its duration is relatively short (for a related view, see Langacker, 1987: 58). Such a sound should tend to stand out because its boundary is "sharper" and more likely to be in short-term memory (because of the sound's short duration). Whether multiple occurrences of a sound are individuated will also depend on having sufficiently long intervals of silence between the sounds. In this case, it should be easier to distinguish the boundary of one occurrence from the boundary of another occurrence.

It also appears that the gerund form of count nouns that name sounds (e.g., yelling, grunting, barking, clicking) systematically refer to nonindividuated sounds and use mass-noun syntax (e.g., 'I heard some yelling in the bedroom,' 'Loud knocking at 3 A.M. woke me up'). Langacker (1987) suggests that any temporal event term formed by adding -*ing* is conceptualized as an ongoing process with the end points of the process outside the scope of reference. Thus, according to Langacker, one reason that these terms are mass nouns is because they refer to unbounded entities. Applying this analysis to sounds, people may conceptualize the referent of a term such as "barking" as an ongoing sound without an onset or offset.

Collective Nouns

A number of singular count nouns are associated with multiple entities and are often called "collective nouns." This diverse group of nouns includes *family, constellation, team, alphabet, jury, bouquet, army, herd, flock, buffet, orchestra, outfit,* and *committee.* Given that a collective noun is a singular count noun, the cognitive individuation hypothesis predicts that people should conceptualize its referent as an individuated entity. Specifically, Wisniewski et al. (2005) suggest that the referent of a collective noun is an *abstract* individual in the sense that it is not a prototypical individual (compare chair, woman), but rather a group of multiple entities conceptualized as a unit or a whole (see also Markman, 1989; Smith and Rizzo, 1982). For example, consider five undergraduates who play basketball for their university—Marcella, Marisa, Carmella, LaChesa, and Esperanza. Although each one can be conceptualized individually as a woman, a student, or a basketball player, in other contexts (e.g., playing together on a basketball court) they are simultaneously conceptualized as a single, more abstract individual—a team.

On this view, the referents of collective nouns have a scope of predication that applies to the entities conceptualized as a whole. That is, central properties apply to the unit or whole of which the multiple entities are composed. They are not predicated individually to each entity that comprises the unit or to arbitrary subsets of these entities. Wisniewski et al. (2005) present an informal analysis of collective nouns that supports this view. For example, a constellation typically refers to a fixed group of stars named after some object, animal, or mythological being that it is shaped like in outline. This property of shape is predicated of the stars as a unit but not of any individual star or of any subset of the stars that composes the constellation. For instance, a single star or arbitrary subset of stars that composes the big dipper does not resemble a dipper in shape. As another example, consider a basketball team. People typically conceptualize a basketball team winning or losing a game rather than a team member or an arbitrary number of team members winning or losing. As a third example, consider a symphony orchestra. People usually conceptualize the conductor and musicians as playing the symphony rather than a single member or arbitrary number of members.

Some research demonstrates factors that induce people to conceptualize multiple entities as an abstract individual. In one experiment, Bloom and Veres (1999) showed that when individual entities in close spatial proximity undergo certain types of synchronized movement, they are perceived as a single unit. Further, on the basis of this synchronized movement, people predicate intention to this unit and not to the entities comprising the unit. Familiar examples of collective nouns that correspond to this type of abstract individual include flocks of birds, herds of animals, and crowds of people moving in unison from or to a location.

There are other factors that lead people to conceptualize multiple entities as a single unit. Wisniewski et al.'s (2005) analysis also revealed that many collective nouns have an internal structure, creating cohesion that may contribute to their conceptualization as an abstract individual. For example, an alphabet is a collection of letters that are linked together in a specific sequence by the successor relation. A suit typically consists of clothing items that are linked together by well-defined spatial relationships (e.g., shirt above pants, tie connected to shirt collar and located in the front middle of shirt). As a final example, consider a family. Many families are characterized as having two adults and some number of children. The adults are linked to each other by a spouse relation and to their children by a parent relation. The children are linked to each other by various brother and sister relations.

In summary, collective nouns appear to refer to abstract individuals—multiple entities that people conceptualize as a single unit or a whole. The central properties of an abstract individual apply to this single unit or whole and not individually to the entities making up the whole or to a subset of entities that are part of that whole. In addition, some collective nouns have an internal structure that links together their constituents, contributing to their conceptualization as a unit.

Pluralia Tantum

A number of plural nouns are associated with multiple objects or objectlike entities and are often called *pluralia tantum*. This diverse group of nouns includes *soapsuds, remains, scissors, groceries, intestines, eyeglasses, supplies,* and *leftovers*. Given that these nouns are grammatically plural, the cognitive individuation hypothesis implies that they should refer to a group of individuated entities (compare plural aggregates, e.g., *keys* and *toothpicks*). However, one typically does not use the singular form of a *plurale tantum* to refer to an entity of which it is composed. For example, it is very unusual to hear someone refer to "a soapsud" or to "a content" (compare "a key" and "a toothpick"). Wisniewski et al. (2005) suggest that *pluralia tantum* refer to *semi-individuated* entities. Specifically, these nouns are plural in order to convey the multiplicity of co-occurring entities that are somewhat distinct. At the same time, their important properties apply to a group of entities as a whole and not individually to the entities making up the group. Thus, their scope of predication is similar to that of collective nouns. There is very little psychological research on the semantics of such terms. Thus, my analysis of *pluralia tantum* will be speculative, drawing on parts of Wierzbicka's (1988) analysis of these terms.

Pluralia tantum can be divided into two broad types: those associated with co-occurring objectlike entities and those associated with co-occurring objects. Some of these terms refer to multiple entities that are objectlike in the sense that each entity is partially but not completely separate from the other entities (e.g., *bleachers, stairs, soapsuds, intestines*). For instance, an objectlike entity that is part of the stairs is somewhat separate from other such entities. At the same time, it is connected to one or two of these entities. Other *pluralia tantum* refer to multiple entities that are objectlike entities in the sense of being degraded versions of previously good individuals. Ruins may include degraded walls, window frames, columns, statues, and so forth, of an ancient building. Remains are often degraded parts of a deceased animal.

A subset of objectlike *pluralia tantum* refer to dual entities and are typically called *summation plurals*. They include *scissors, eyeglasses, pants, goggles, pliers, binoculars, tweezers, earmuffs,* and *headphones*. According to Wierzbicka (1988: 515), these terms "designate objects which have two identical parts fulfilling the same function within the whole." The two parts are partially separated from another so that one can think of them as being objectlike. However, they are also joined together as a necessary condition for the object to achieve its function. For example, one cannot cut things with just one of the identical objectlike entities of scissors, nor keep the ears warm with just one of the objectlike entities of earmuffs. On this view, the important properties of summation plurals apply to its entities as a unit or as a whole. Correspondingly, the entities (i.e., the two identical parts) lack these important properties, and one does not refer to either of the parts with the singular form of the category (e.g., a scissor). At the same time, the co-occurring parts look somewhat like distinct objects. Thus, members of these categories are named with plural nouns.[2]

The hypothesized conceptual basis for these *pluralia tantum* is reinforced by contrasting them with other dual entities that are not summation plurals but rather refer to two distinct individuals. They include *ears, eyes, gloves, socks, shoes, hands,* and *legs*. Although these terms refer to dual entities as do summation plurals, they have a different scope of predication. Specifically, important properties of these dual entities apply to each entity making up the duo. Thus, people can also use a singular count noun to refer to one of the entities of the duo, unlike summation plurals. There appears to be evidence that is consistent with this view. For example, consider *ears* and *eyes*. Although we primarily use both ears to hear and both eyes to see, we can achieve these functions with just one ear or just one eye (compare trying to cut something with one part of scissors, a *plurale tantum*). In addition, we often interact in the same way with both ears and eyes, but the interactions are temporally separate. For instance, one puts a pair of earrings on the ears one ear at a time and puts eye drops in the eyes one eye at a time. In contrast, important interactions with entities named by summation plurals apply to both entities simultaneously (e.g., we

---

2. Some languages (e.g., Dutch and German) use singular nouns to refer to objects with identical pairs of objectlike parts (Bock et al., 2001). One might expect such language differences given that the status of these objectlike parts as individuals is ambiguous.

simultaneously attach the two objectlike entities of the earmuffs to the ears). These differences in mode of interaction may give rise to the perception of important predicates applying individually or as a whole, respectively. As another example, many important functions of hands can be carried out with either hand (e.g., waving goodbye, picking something up, knocking on a door).

The second type of *pluralia tantum* typically consists of groups of objects rather than objectlike entities. These object-based *pluralia tantum* include *leftovers, groceries, contents, refreshments, goods, goodies, spoils, supplies, odds and ends, belongings*, and *valuables*. Wierzbicka (1988: 540) suggests that these *pluralia tantum* refer to "a number of different things" that "are given jointly a common name...because they are all in the same place, at the same time, and...they are in that place at that time for the same reason."

To illustrate these characteristics consider the object-based *pluralia tantum groceries* and *contents*. Groceries are typically found together in a grocery cart, having been put there by someone who is shopping. Later, a grocery bagger places them into bags, and the shopper transports the bags home. Furthermore, the entities that comprise groceries can vary greatly— one could refer to a bag containing apples, paper towels, dog biscuits, shrimp, a newspaper, milk, and contact lens solution as groceries (if these items had been purchased together at the supermarket). As another example, contents exist together (e.g., in a closet) and were put there by someone. One could refer to a collection of virtually any group of entities as the contents of a closet as long as they existed together in the closet.

Like objectlike *pluralia tantum*, people may conceptualize the important properties of object-based *pluralia tantum* as applying to the collection of entities as a whole and not individually to each entity making up the collection. For instance, one pays an amount of money that covers the cost of all the groceries together, transports them together, carries them together into one's residence (which are usually contained together in bags), and so on.

Given this analysis it is instructive to summarize the conceptual differences and similarities among *pluralia tantum*, aggregates, mass-noun superordinates, and collective nouns. *Pluralia tantum* are similar to aggregates, mass-noun superordinates, and collective nouns in that their referents are associated with multiple entities. At the same time, they are grammatically different from these other language terms, suggesting that they are conceptually different as well.

Pluralia tantum contrast with count-noun aggregates in scope of predication. Important properties of *pluralia tantum* apply to multiple entities as a whole, whereas they apply to each constituent of the count-noun aggregate. *Pluralia tantum* and mass-noun aggregates appear to differ in two ways. First, the constituents of *pluralia tantum* are perceived more as distinct entities than those of mass-noun aggregates. (Recall that the constituents composing mass-noun aggregates are more difficult to perceptually distinguish than those of count-noun aggregates.) Second, *pluralia tantum* and mass-noun aggregates may differ in scope of predication. Important properties apply to *pluralia tantum* as a whole but not to arbitrary subsets of their constituents. For example, one uses the term *groceries* to denote a particular collection of items purchased on a specific occasion. Thus, one probably would not consider an arbitrary subset of these items to be groceries. As another example, an important property of STAIRS is "allows one to go from one floor to another floor." This property does not apply to an arbitrary subset of these stairs. In contrast, as previously discussed, important properties of mass-noun aggregates can apply to arbitrary sets of their constituents. For example, an important property of rice is that 'it is consumed,' and people can consume varying amounts of rice or a person can eat varying amounts on different occasions. The contrast in scope of predication between *pluralia tantum* and mass-noun aggregates may also apply to *pluralia tantum* and mass-noun superordinates. That is, important properties apply to *pluralia tantum* as a whole but not to arbitrary subsets of their constituents, as they appear to do with mass-noun superordinates.

Finally, *pluralia tantum* are similar in scope of predication to collective nouns (i.e., important properties apply to multiple entities as a whole). However, they lack the internal structure and relations between entities that give rise to the conceptualization of those entities as a single unit (compare *groceries, contents,* and *leftovers* vs. *family, constellation,* and *alphabet*). For example, the stars that compose a constellation have well-defined, psychologically fixed spatial relations and distances between them, leading to the perception of a single entity. In contrast, such relationships characterize a shopping cart or bag of groceries to a much lesser degree.

Summary

In this section, I have presented evidence from a variety of areas that supports the cognitive individuation hypothesis. This evidence suggests that people tend to use count nouns when they have distinct individuals in mind but mass nouns when they have nonindividuated entities in mind. Thus, there is a systematic relationship between the use of count and mass nouns and how we conceptualize the referents of these terms. The evidence also shows that how people interact with entities strongly affects whether they conceptualize entities as individuated or nonindividuated. Some of the evidence also reveals a conceptual basis for the count–mass noun distinction among language terms for which a conceptual basis was thought not to exist (e.g., superordinate categories).

Beyond the Cognitive Individuation Hypothesis

Although there is strong evidence for a systematic relationship between conceptualization and count–mass noun syntax, the relationship is not perfect. For example, in the studies of aggregates described previously, Middleton et al. (2004) found that 19 of the 192 aggregates contradicted their predictions. That is, the perceptual distinguishability and interaction ratings predicted that a count-noun aggregate should have mass-noun syntax (or vice versa for a mass-noun aggregate).

Some of these exceptions may be explained by other factors that affect whether entities are conceptualized as individuated. Thus, they may not be true exceptions to the cognitive individuation hypothesis. For example, four of the count-noun exceptions (*fleas, bacteria, maggots,* and *lice*) were animate. Even though their perceptual distinguishability and interaction ratings predict that they should have mass-noun syntax, animacy may also cause people to individuate the elements of an aggregate. (There do not appear to be any aggregates of animate entities that are nonindividuated.)

Nevertheless, Middleton et al. (2004) identified nine exceptions that could not be explained on the basis of other factors relevant to individuation (*fibers, aspirin, Advil, Tylenol, firewood, asparagus, bacon, money* [in a wallet], and *candy*). For example, the mass-noun exceptions *aspirin* and *bacon* appear to have retained the syntax that reflects how their original referents were conceptualized. At one time, they referred to prototypical nonindividuated entities. In particular, aspirin used to be manufactured and administered as a powder, and bacon used to refer to fresh pig meat. Today, though, *aspirin* are tablets and people typically interact with them one or a few at a time (as we found in the interaction rating study). Still, people call them *aspirin* (compare *vitamins*). Likewise, *bacon* now refers to a cured slab that is presliced, and people typically eat one or a few at a time (as we found in the interaction rating task). Yet, people call them bacon (compare *french fries*). It is unclear why English speakers have retained the mass-noun syntax of these entities. As another example, Paul Bloom, a major proponent of the cognitive individuation hypothesis, has said: "I have no doubt that I think of a piece of toast as a singular individual, but—due to a quirk of English—I have to talk about it using the word 'toast,' a mass noun. So I ask you, 'Do you want more toast?' while thinking of a singular entity."[3]

Importantly, the syntax of a language may serve multiple communicative functions that compete with each other. Because speakers may wish to convey some other important aspect of an entity, they may refer to that entity using syntax that conflicts with the *individuation function* (i.e., the tendency to use count-noun syntax to refer to an individuated entity and mass-noun syntax to refer to a nonindividuated entity).

Wisniewski et al. (2003) noted that food is a domain in which communicating about other aspects of an entity takes precedence over the individuation function. Various kinds of foods undergo transformations that change their status from individuals to nonindividuated entities as a result

---

3. Paul Bloom (personal communication, July 2002).

of chopping, dicing, mashing, scrambling, and so forth. For example, people dice tomatoes, chop peanuts, scramble eggs, and mash potatoes. These transformations destroy the integrity of the individuals and produce substances. Yet, the syntax of the names for these entities does not reflect this change to a nonindividuated entity. For example, American English speakers refer to potatoes that have been mashed as "mashed *potatoes*" and to eggs that have been scrambled as "scrambled eggs." However, it doubtful that speakers conceptualize them as multiple individuals, as the syntax would imply.

Evidently, speakers base their names for these substances on information that characterizes how the substances originated. Specifically, the names *mashed potatoes* and *scrambled eggs* reflect the type of transformation and type of entity to which the transformation was applied. For example, one may refer to a yellowish, fluffy, edible substance with the plural count-noun phrase "scrambled eggs" because the scrambling was applied to multiple individuals (i.e., eggs). Thus, the name for the resulting substance inherits the count—mass noun syntax of the entity that was transformed to produce that substance. People also may be reluctant to use the mass-noun phrase "scrambled egg" because it would incorrectly imply that the scrambling was applied to a substance.

Wisniewski et al. (2003) report results of an internet search that supports this view. They found that the percentage of hits for "scrambled eggs" or "some scrambled eggs" compared to "scrambled egg" or "some scrambled egg" was 89% (out of 38,374 hits). Likewise, the percentage of hits for "mashed potatoes" or "some mashed potatoes" compared to "mashed potato" or "some mashed potato" was also 89% (out of 116,620 hits). Presumably, in coining the phrases "mashed potatoes" and "scrambled eggs," information about the origin of and process that produced the substance was more important to convey about the entity than its individuation status. Wisniewski et al. (2003) also found that the names of many fruit or vegetable substances provide information about the type of transformation and type of entity (i.e., multiple individuals) to which the transformation was applied. As a result, the name for the substance inherited the count-noun syntax of the multiple individuals that were transformed into that substance. For example, phrases such as *chopped radishes, diced radishes*, and *mashed radishes*, were much more frequent than *chopped radish, diced radish*, and *mashed radish*, even though the latter phrases refer to foods transformed into substances.

There are also nonfood cases in which communicating about other aspects of an entity takes precedence over the individuation function. For example, speakers use the count-noun phrases *eye drops* and *artificial tears* to refer to a variety of liquids contained in small bottles (and manufacturers have labeled the bottles with these names). When referring to one of these liquids with a plural count noun it is doubtful that people are thinking of the liquid as a collection of individual drops of liquid or tears. Instead, these names have dual reference. They refer to actual drops of liquid or tearlike liquid as well to the liquid contents of small bottles. Evidently, the names reflect the important functions of these liquids (i.e., placing *drops* of liquid into the eyes to correct eye problems). Naming the actual liquid based on function took precedence over naming based on the individuation function.

The phrase *pine needles* may be an example of naming based on resemblance. The phrase was coined to refer to the leaves of a pine tree that resemble needles (they are also prickly to some extent like needles). Actual needles are good examples of individuals. People frequently interact with them one at a time and the important properties of needles are predicated of each individual needle (e.g., sewing a button, drawing blood, administering a vaccine). In contrast, pine needles are a good example of a nonindividuated entity. They occur as multiple entities found on the ground in very close proximity and are difficult to perceptually distinguish. People also interact with multiple pine needles at a time (when using them as mulch). Further, individual pine needles are not seen falling from a tree as is the case with other leaves. Theses characteristics are associated with aggregates consisting of nonindividuated entities. Evidently, whoever named pine needles *pine needles* was struck by their resemblance to needles and retained the count syntax of needle at the expense of the individuation function.

As another example, of the interaction between multiple communicative functions, Malt et al. (1999) emphasize that a person may accept a name for an entity because it facilitates communication and thus focus less on its similarity to other entities in that category. For example, someone may adopt a product name provided by an advertiser because it is known to and used by others. With respect to the count—mass noun distinction, a recent example is *Egg Beaters*. This name refers to processed egg white *substance*. Evidently, manufacturers based its name on function, which took precedence over naming based on the individuation function. (It is called *Egg Beaters* because it "beats" the cholesterol problem associated with eggs.)

The Role of Individuation in Cognition

The conceptualization of an entity as individuated or nonindividuated appears to be a very basic process that pervades all of cognition. For example, counting requires that one individuate some number of entities. Categorizing entities requires that they be individuated or treated as nonindividuated, depending on the type of entity. For example, categorizing some entity as a cat requires that one conceptualize the cat as a distinct individual, and to categorize something as sand one must conceptualize that sand as nonindividuated.

The process of individuation may also affect memory and attention. For example, consider differences in how people might attend to a nonindividuated entity such as a substance versus an individuated entity such as a physical object. Construing something as a nonindividuated substance suggests that its texture and color are important and that shape is irrelevant. In contrast, construing something as an individuated physical object suggests that shape is important and that texture and color are less relevant (see also Imai and Gentner, 1997; Soja et al., 1991). As a result, people might attend to different features in the two cases that in turn could affect memory for these features.

Individuation also affects the types of inferences that people draw about the environment. For example, the inferences that one makes when hearing "some chicken" versus "a chicken" can be dramatically different. The former can refer to a nonindividuated substancelike entity that is likely to be found in a kitchen, have an expiration date, and be cooked and eaten. The latter can refer to a live bird and hence is likely to be found on a farm, have wings, fly, be alive, and so forth. These different inferences depend on specific knowledge about chickens. However, it is the distinction between a count noun and a mass noun that indicates whether the chicken refers to an individuated or nonindividuated entity, which in turn enables people to access the appropriate information. Sometimes differences in inferences can be quite subtle. For example, if you heard someone say, "I heard some noise in the kitchen" versus "I heard a noise," you might infer that the former was of longer duration. As another example, the phrase "too much curiosity" typically refers to an amount of curiosity, whereas "a curiosity" refers to a cause of the curiosity.

Summary and Conclusions

Scholars have long debated why English and other languages make a grammatical distinction between count and mass nouns. These debaters fall into two camps: one maintains that the distinction is primarily an arbitrary convention of language, and the other maintains that the distinction is primarily conceptually motivated. They account for any apparent arbitrariness by postulating that the referents of count and mass nouns cannot always be identified on the basis of their obvious perceptual characteristics. Rather, speakers flexibly construe the referents of count and mass nouns as individuated and nonindividuated entities, respectively. I have described a range of evidence that supports this conceptually motivated distinction. At the same time, this view must be qualified. Other functions of language can interact with the individuation function conveyed by count- and mass-noun syntax. Future work needs to specify the nature of these interactions and address other domains that make a grammatical distinction between count and mass nouns (e.g., abstract concepts, sounds).

ACKNOWLEDGMENTS

I thank an anonymous reviewer for trenchant and detailed comments on a previous draft of this chapter, and Bob Dylan for providing some of the inspiration for this work. This research was supported by National Science Foundation grant BCS-9975198.

REFERENCES

Bloom, P. (1990). Semantic structure and language development. Unpublished doctoral dissertation, Massachusetts Institute of Technology, Cambridge.

Bloom, P. (1996). Possible individuals in language and cognition. *Psychological Science, 5,* 3, 90–94.

Bloom, P., and Veres, C. (1999). The perceived intentionality of groups. *Cognition, 71,* B1–B9.

Bloomfield, L. (1933). *Language.* New York: Holt and Co., 266–268.

Bock, K., Eberhard, K.M., Cutting, J.C., Meyer, A.S., and Schriefers, H. (2001). Some attractions of verb agreement. *Cognitive Psychology, 43,* 83–128.

Gleason, H.A. (1969). *An introduction to descriptive linguistics.* London: Holt, Rinehart, and Winston.

Gordon, P. (1985). Evaluating the semantic categories hypothesis: The case of the count/mass distinction. *Cognition, 20,* 209–242.

Imai, M., and Gentner, D. (1997). A cross-linguistic study of early word meaning: Universal ontology and linguistic influence. *Cognition, 62,* 169–200.

Langacker, R.W. (1987). Nouns and verbs. *Language, 63,* 53–94.

Lucy, J.A. (1992). *Grammatical categories and cognition.* Cambridge: Cambridge University Press.

Malt, B.C., Sloman, S.A., Gennari, S., Shi, M., and Wang, Y. (1999). Knowing versus naming: Similarity and the linguistic categorization of artifacts. *Journal of Memory and Language, 40,* 230–262.

Markman, E.M. (1985). Why superordinate category terms can be mass nouns. *Cognition, 19,* 31–53.

Markman, E.M. (1989). *Categorization and naming in children.* Cambridge, Mass.: MIT Press.

McCawley, J. (1975). Lexicography and the count-mass distinction. *Berkeley Linguistic Society, 1,* 314–321.

McPherson, L.P. (1991). A little goes a long way: Evidence for a perceptual basis of learning for the noun categories COUNT and MASS. *Journal of Child and Language, 18,* 315–338.

Middleton, E.L., Wisniewski, E.J., Trindel, K.A., and Imai, M. (2004). Separating the chaff from the oats: Evidence for a conceptual distinction between count noun and mass noun aggregates. *Journal of Memory and Language, 50,* 4, 371–394.

Murphy, G.L., and Wisniewski, E.J. (1989). Categorizing objects in isolation and in scenes: What a superordinate is good for. *Journal of Experimental Psychology: Learning, Memory, and Cognition, 15,* 572–586.

Palmer, F.R. (1971). *Grammar.* Harmondsworth: Penquin Books, 34–35.

Quine, W.V. (1960). *Word and object.* Cambridge, MA: MIT Press.

Sloman, S.A., Love, B.C., and Ahn, W. (1998). Feature centrality and conceptual coherence. *Cognitive Science, 22,* 189–228.

Smith, L. B., and Rizzo, T.A. (1982). Children's understanding of the referential properties of collective and class nouns. *Child Development, 53,* 245–257.

Soja, N.N., Carey, S., and Spelke, E.S. (1991). Ontological categories guide young children's inductions of word meaning: Object terms and substance terms. *Cognition, 38,* 179–211.

Ware, R. (1979). Some bits and pieces. In F. Pelletier (Ed.), *Mass terms: Some philosophical problems.* Dordrecht: Reidel, 15–29.

Whorf, B.L. (1962). *Language, thought, and reality.* Ed. John Carroll. New York: John Wiley and Son, 140–141.

Wierzbicka, A. (1988). *The semantics of grammar.* Amsterdam: John Benjamins.

Wisniewski, E.J., and Imai, M., and Casey, L. (1996). On the equivalence of superordinate concepts. *Cognition, 60,* 269–298.

Wisniewski, E.J., Lamb, C.A., and Middleton, E.L. (2003). On the conceptual basis for the count and mass noun distinction. *Language and Cognitive Processes, 18,* 5/6, 583–624.

Wisniewski, E.J., Clancy, E., and Tillman, R. (2005). On different types of categories. In W. Ahn, R.L. Goldstone, B.C. Love, A.B. Markman, and P. Wolff (Eds.), *Categorization inside and outside of the laboratory: A festschrift in honor of Douglas Medin.* Washington, DC: American Psychological Association, 103–126.

10

Count Nouns, Sortal Concepts, and the Nature of Early Words

FEI XU

Early words consist of mostly count nouns. A subset of our concepts, *sortals*, underpins our representations of count nouns. A sortal concept is a concept that provides principles of individuation and principles of identity (Hirsch, 1982; Macnamara, 1986; Wiggins, 1980). To answer the question "how many?" we need to specify "how many what." If we are interested in counting the number of things in a room, we would receive different answers by asking "how many *tables*," "how many *chairs*," or "how many *legs*." Similarly, to answer the question "is it the same," we need to specify "the same what." A person may not be the "the same *baby*" as she was 17 years ago, but she may still be "the same *person*." Our identity criteria are also sortal-relative in the sense that the same property difference may or may not indicate a change of identity, depending on the kind of object in question (e.g., a change in size and color indicates a change in identity for a chair but not necessarily for a plant). Sortals are the concepts that provide the criteria to enumerate and track identity over time and they are lexicalized as count nouns in languages that make the count—mass distinction (Baker, 2003; Hirsch, 1982; Macnamara, 1986; Wiggins, 1980). All concepts provide principles of application (i.e., specifying what falls under the concept), but not all concepts provide principles of individuation and identity. Consider the concept RED. We cannot count "the red" in a room, unless we specify a sortal, for example, 'red *shirts*,' 'red *lights*,' or 'red-*heads*.' We also cannot count "the good" but we can count the number of 'good *people*,' 'good *thieves*,' or 'good *knives*.' Similarly, we cannot ask whether something is 'the same red' or 'the same good' unless we mean 'the same red shirt' or 'the same good thief.' The interpretation of *red* or *good* differs drastically depending on whether the noun is 'shirt,' 'head,' 'person,' or 'thief' (for a more nuanced discussion of adjective meanings, see Partee, 1990). Generally speaking the interpretation of predicates (be they adjectives, verbs, or other grammatical classes) depends on the noun. Mass nouns such as *sand* and *water* differ from count nouns in that they do not provide principles of individuation and identity in a straightforward way. Some have suggested that *portions of substance* provide principles of individuation and identity (e.g., Hirsch, 1982). For example, we can distinguish one pile of sand from two, and three glasses of water from five glasses of water.

This chapter has two main sections. In section 1, I review a body of research investigating how representations of sortal concepts develop in infancy and how learning count nouns may play a causal role in constructing these concepts. In section 2, I suggest that the work on sortal concepts as well as other related research argue against the traditional view of early word learning, namely, that early words are fundamentally different in character from later words.

1. Object Individuation and Sortal Concepts

Object individuation is the process by which one establishes the number of distinct objects in an event. In particular, it is concerned with the process whereby an object is seen at time 1 and an object is seen on time 2, and the question arises as to whether they are the same object seen on two different occasions or two distinct objects. As mentioned above, a sortal concept is a concept that provides principles of individuation and principles of identity.

This section focuses on the developmental origin of the representation of sortal concepts. I use the criteria by which children and adults individuate objects as a means for investigating when children begin to represent sortal concepts. For adults, at least three sources of information are regularly employed in individuating objects: spatiotemporal information, property (or featural) information, and sortal information. The use of spatiotemporal information includes generalizations such as objects travel on spatiotemporally continuous paths, or two objects cannot occupy the same space at the same time. The use of property information includes generalizations such as objects do not usually change shape, size, or color. The use of sortal information includes principles such as objects do not change kind membership; thus, if an object seen at time 1 falls under one sortal concept and an object seen at time 2 falls under another sortal concept, they must be two distinct objects. Furthermore, property information is sortal-relative such that property differences are weighted differently depending on the kind of object under consideration. Note that for adults, property information is sortal-relative for known kinds (and kinds that can be easily assimilated to known kinds), but we also use property information in a domain-general way (e.g., objects tend to have regular shapes and don't usually change color).

The developmental evidence I review suggests that young infants use spatiotemporal information for object individuation, but it is only later that they begin to use property information to do so. It is later still in development that they begin to use sortal information for object individuation, and the emergence of this ability coincides with when infants start to comprehend their first words for objects. I argue that (1) it may be adaptive to rely on spatiotemporal information for object individuation early on, and the use of property information is secondary; (2) the use of sortal information may require conceptual change on the part of the infant; and (3) language learning may play an important role in inducing such conceptual reorganization.

Using the violation-of-expectancy looking time methodology (Spelke, 1985), a number of studies have shown that infants as young as 2.5–4 months represent persisting objects, and they employ spatiotemporal information to determine how many objects are in an event. In a seminal study, Spelke et al. (1995) showed that if objects appear to have traveled on spatiotemporally discontinuous paths (figure 10.1), the infants posited two distinct objects in the event. That is, they looked longer at the unexpected outcome of one object than the expected outcome of two objects. Other laboratories have replicated and extended these findings (Aguiar and Baillargeon, 1999; Simon et al., 1995; Spelke, 1990; Wynn, 1992). These studies have been taken as evidence that infants represent the sortal concept OBJECT (Xu, 1997, 2005; Xu and Carey, 1996), although it is a matter of controversy whether OBJECT is a full-fledged sortal (e.g., Hirsch, 1997; Wiggins, 1997).

What about the use of property or sortal information for object individuation? Can infants use object properties, such as the shape, size, and color of objects, or sortal concepts such as DUCK or BALL in deciding how many objects are in an event? Some studies suggest that it is not until twelve months of age that infants are able to use property or sortal-kind information in the service of object individuation (Xu and Carey, 1996). In these experiments, infants saw an object (e.g., a toy duck) appear from behind an opaque screen then return behind it. Then they saw an object (e.g., a ball) appear from behind the same screen then return behind it (figure 10.2). This event was repeated several times. Then the screen was removed to reveal either two objects (the duck *and* the ball, the expected outcome) or just one of the two objects (the duck *or* the ball, the unexpected outcome). Infants' looking times for these outcomes were recorded. At ten months, infants did not look longer at the unexpected outcome of one object; their looking times were not different from their baseline preference for two objects. At twelve months, however, infants looked longer at the unexpected outcome of one

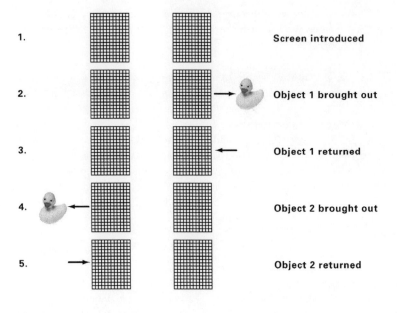

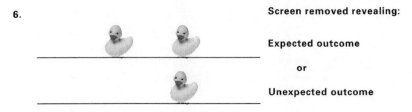

FIGURE 10.1 Discontinuous condition

object. Xu and Carey (1996) suggested that ten-month-old infants did not use property or sortal information to establish a representation of two distinct objects, whereas twelve-month-old infants did. Importantly, control experiments showed that the infants had encoded the perceptual differences between the objects—they habituated faster to a sequence of duck, duck, duck, duck relative to a sequence of duck, ball, duck, ball—but they failed to use these differences to compute the number of objects in the event. Other laboratories have replicated and extended these findings using looking time as well as manual search measures (Bonatti et al., 2002; Krøjgaard, 2000; Rivera and Zawayden, 2007; Van de Walle et al., 2000; Wilcox and Baillargeon, 1998a; Xu et al., 1999, 2004).

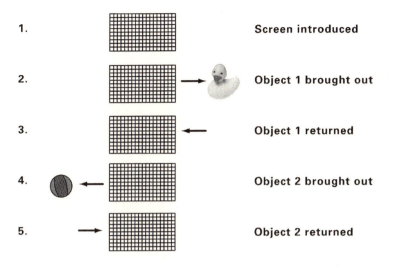

FIGURE 10.2 Property/kind condition

Crucial for our investigation of sortal concepts, we needed to ask whether twelve-month-old infants' success was based on their representations of sortal concepts (e.g., the difference between a DUCK and a BALL), or whether it was based on property differences (e.g., the difference between a yellow, irregularly shape object and a round, red and green object). Xu et al. (2004) conducted a series of experiments to tease these possibilities apart. The same is-it-one-or-two task was used, but sometimes the infants saw two objects that differed only in color (e.g., a red ball and a green ball), size (e.g., a small ball and a large ball), or a combination of these features. In each case, although the infants had encoded the perceptual differences between the objects, they failed to use these property differences to establish a representation of two objects. Shape contrasts were also investigated: when difference in shape signaled a sortal contrast (e.g., a plastic, yellow CUP and a plastic, yellow BOTTLE), infants succeeded. When a similarly salient shape difference (measured by habituation rate) did not signal a sortal contrast (e.g., a plastic, yellow regular CUP and a plastic, yellow sippy CUP), infants again failed to infer the presence of two objects. Thus, sortal representations appear to be the basis of success at twelve months. Parallel results were found using doll-heads (Bonatti et al., 2002) at ten months—with an ontological distinction such as *human* versus *nonhuman*, infants succeeded on the individuation task with the contrast a doll head versus a cup, but they failed with a female doll versus a male doll. Taken together, these studies suggest that representations of sortal concepts begin to emerge at around ten to twelve months.

It is controversial, however, whether infants younger than twelve months are able to use property information for object individuation. The evidence is mixed, depending on the task demands and the specifics of the dependent measures. Using simplified versions of the object individuation task, some evidence suggests that at around ten months (or even younger) infants are able to use property information for object individuation (looking-time measure: Wilcox, 1999; Wilcox and Baillargeon, 1998a, 1998b; Wilcox and Chapa, 2004; Wilcox and Schweinle, 2002; manual search measure: Xu and Baker, 2005), but some have provided alternative interpretations for these results (e.g., Xu, 2005; Xu et al., 2004).

How do infants acquire sortal concepts? It seems a suspicious coincidence that most infants begin to comprehend words for basic-level object categories at around ten to twelve months, and they also begin to represent (basic-level) sortal concepts such as DUCK and BALL at around the same time. My colleagues and I explored the hypothesis that perhaps learning words for object categories plays an important role in acquiring sortal concepts (Xu, 2002; Xu et al., 2005). Nine-month-old infants were presented with the is-it-one-or-two task described above. This time, however, when each object emerged from behind the screen, the experimenter said, "Look, a duck!" or "Look, a ball!" With just a few repetitions of these labels, infants looked longer at the unexpected outcome of one object than the expected outcome of two objects on the test trials. Infants also succeeded when two unfamiliar objects and nonsense words were used. However, they failed when both objects were labeled "a toy," or when two distinct tones, sounds, or emotional expressions were provided. We suggested that infants expect words (count nouns) to refer to sortals and that the use of two distinct labels signaled to the infant that two kinds of objects were presented in the event, so two objects must be behind the screen. Other laboratories have replicated and extended these results (Fineberg, 2003; Rivera and Zawaydeh, 2007).

But perhaps the words simply provided the 9-month-old infants with a mnemonic on-line during the experiment, with no lasting effects for representations of sortal concepts in the real world. Some evidence suggests that this is not the case; word learning may be integral to acquiring these concepts. In two studies, parents of ten- and eleven-month-old infants were asked to report on their infants' word comprehension for a set of highly familiar objects. When these objects were used in the is-it-one-or-two object task without labeling, the results showed that infants who knew both words for the objects used in the task succeeded but those who did not know the words failed (Rivera and Zawaydeh, 2007; Xu and Carey, 1996). Another study asked whether labeling alone could guide the process of establishing representations of distinct objects. Using a manual search method, twelve-month-old infants were shown to be able to use the presence of labels to determine how many objects were in a box whose content was invisible to them (Xu et al., 2005). When infants heard the content of the box labeled with two words, they expected to find two objects inside; when they heard just one word repeated, they expected to find only one object inside the box. This effect appeared to be language-specific since infants did not expect to find two objects when two emotional expressions were used.

Do infants, like adults, expect two distinct labels to refer to two kinds of objects and not just two individual objects? In some recent studies, nine-month-old infants were tested to see if they had the same expectation about words (Dewar and Xu, 2007). Infants were first shown two possible outcomes, either two identical objects or two objects differing in shape, color, and surface pattern. Then the infants were given linguistic information about the content of the box using either two labels or one label. Looking-time results showed that infants expected to see the different object outcome when they heard two labels and the identical object outcome when they heard just one label. Furthermore, follow-up experiments showed that infants' expectations were not satisfied by just any difference between the two objects: they expected two different labels to map onto two differently *shaped* objects. Color alone was not sufficient to satisfy infants' expectations. Since shape is a perceptual dimension often correlated with kind membership (at least for the kinds of objects we used in the experiments) and color is not, it appears that even nine-month-old infants expect distinct count nouns to map onto distinct kinds of objects, not just individual objects.

In sum, these studies suggest that the criteria by which infants individuate objects change over the course of the first year of life. Spatiotemporal information may be primary early on, and property and sortal information is employed later. This developmental trajectory may be adaptive since no physical objects violate spatiotemporal continuity whereas property or sortal information depends on learning about different sorts of objects in the real world. By the end of the first year of life, infants begin to represent (basic-level) sortal concepts such as DUCK and BALL. This conceptual change allows the infants to see the world in terms of kinds of things and not just objects with various properties. From this point on infants, like adults, presumably begin to organize property information around sortals—the same property difference may or may not indicate a change of object identity depending on the kind of thing it is. Furthermore, learning words for objects may play an important role in the acquisition of sortal concepts (the concept of PERSON/HUMAN may be an exception; see Bonatti et al., 2002).

Other aspects of conceptual development have also been shown to be influenced by linguistic information. In several categorization studies, Waxman and her colleagues found that the presence of a count noun facilitated categorization in infants as young as nine months, and the facilitation effects are linked to grammatical classes by about thirteen months (Balaban and Waxman, 1996; Waxman, 1999; Waxman and Braun, 2005; Waxman and Markow, 1995). In several inductive inference studies, Graham and her colleagues found that providing a count noun label allowed thirteen- and eighteen-month-old infants to make inferences about nonobvious properties of objects (e.g., squeeze an object to make a sound) above and beyond perceptual similarity (Graham et al., 2004; Welder and Graham, 2001). These labeling effects become more fine-tuned according to grammatical classes (noun vs. adjective) by eighteen months (Joshi and Xu, 2006). These studies converge with the results of the object individuation studies: infants expect count nouns to map onto kinds of objects at the beginning of word learning, and this expectation leads them to use labeling as a source of information for identifying the kinds in their environment. The labeling event ("Look, a rabbit!") informs the infant that she should set up a mental symbol that represents a sortal concept. Furthermore, the kind of object this word refers to has an essence that determines its internal and surface properties. In this sense, words may be "essence-placeholders" for young children (Gelman, 2003; Medin and Ortony, 1989; Xu, 2002, 2005).

2. Early Word Learning

In this section, I suggest that the research on object individuation, categorization, and inductive inference in infancy also bears on how we think about the nature of early words. In particular, I argue that these data support the view that even children's earliest words for objects refer to kinds, that is, the principle of generalization. I also touch on two other characterizations of words—the principle of reference and the principle of conventionality, but the discussion is short and more speculative since a detailed discussion is beyond the scope of this chapter.

Many studies have tried to teach infants between ten and eighteen months a new word (a count noun) for a novel object in a laboratory setting. The results are mixed. Under some circumstances thirteen- to fourteen-month-old infants succeed in learning a new word rather quickly with just a few exposures (e.g., Waxman and Booth, 2001; Woodward et al., 1994). Other times it takes a lot of repetitions for infants at fourteen or fifteen months of age to establish word–object mappings (e.g., Hollich et al., 2000; Schaeffer and Plunkett, 1998; Werker et al., 1998). Some studies showed successful generalization to objects of the same kind at thirteen or fourteen months (e.g., Waxman and Booth, 2001; Woodward et al., 1994), while others found it difficult to show generalization of a new word until 19 months of age (e.g., Hollich et al., 2000).

Many researchers have suggested that the slow rate of vocabulary development early on reflects a different process of learning. A few standard facts can be found in any textbook on language development and developmental psychology in general. First, most infants begin to comprehend single words at around ten to twelve months. Production follows, but individual children vary a great deal in when they reliably produce words. Second, many of the early words are names for objects. Third, early word learning appears to be very slow and effortful. Vocabulary grows very slowly, and infants may show comprehension of a word intermittently. Fourth, for many children, the rate of vocabulary learning takes off at around eighteen months, that is, the "vocabulary spurt" (although see Bloom, 2000; Ganger and Brent, 2004).

Based on this characterization of early words, many have suggested that early words are also different in character compared with later words. Three characteristics have been suggested: (1) early words are context-bound, (2) early words are the result of associative learning, and (3) early words are idiosyncratic (e.g., Dromi, 1987; Golinkoff et al., 1994; Lock, 1980; McShane, 1979). In contrast, the more mature lexicon shows rather different characteristics. Three characteristics have been suggested for the count nouns in the lexicon: (1) principle of generalization: count nouns refer to kinds/sortals, (2) principle of reference: count nouns refer to, and are not merely associated with, objects, and (3) principle of conventionality: word meanings are shared by members of a particular linguistic community (e.g., Baldwin, 1993; Bloom, 2000; Clark, 1983; Diesendruck and Markson, 2001; Landau et al., 1998; Markman, 1989).

Is there evidence that early words/count nouns are fundamentally different from later words? The literature provides mixed results. In a seminal study, Huttenlocher and Smiley (1987) analyzed the spontaneous speech of very young children who were just beginning to learn their first words. They found no evidence that children's early words were context-bound; that is, young children do not use the word *car* only when looking out the window as if the meaning of the word includes the context in which it had been learned. They also found very little overgeneralization in the comprehension data; that is, children did not use say the word *dog* to refer to all animals, but rather used it appropriately (see also Huttenlocher, 1974; Macnamara, 1982). Other studies have found more overgeneralization of early count nouns (e.g., Clark, 1983; Gelman et al., 1998; Naigles and Gelman, 1995; Nelson, 1973).

In the studies on early conceptual development reviewed above, we see a different picture: infants are very competent in using linguistic information for other cognitive tasks such as categorization, individuation, and inductive inference as early as nine to thirteen months. This seems to pose a puzzle: on the one hand, learning specific mappings between words and objects is difficult for infants younger than eighteen months; on the other hand, children of the same age are already able to use linguistic information to facilitate their conceptual development. How can we reconcile these findings? Here is a suggestion: success in the conceptual tasks only requires the infants to use *abstract properties* of words, namely, the expectation that words map onto sortals and sortal concepts support categorization, individuation, and inductive inference. Success in learning new words for objects, however, requires *specific mappings* that include learning the phonological form of the word, the appearance of the object, and the mapping between the two. This process requires attentional resources that may be scarce in infants.

Principle of Generalization

For older children and adults, count nouns refer to kinds of individuals. What are some cognitive functions of kind representations? There are at least three. Kind concepts allow us to categorize objects. If several objects are all called "blicket," we know that they are members of the same kind/category (note exceptions such as "bat"). Kind concepts also allow us to individuate objects. If an object is seen at time 1 and it is called a "blicket," and an object is seen at time 2 and it is called a "fep," we infer that since they are different kinds they also must be two numerically distinct objects (note that this is likely to be a default assumption since there are cases such as "a dog" vs. "a pet" or "an animal" vs. "a poodle"). Lastly, kind concepts support inductive inference. If we know that blickets make a distinct sound, we infer that the next blicket we encounter would make the same sound.

A growing body of work on early conceptual development has investigated whether kind concepts support these three cognitive functions in infants. The emphasis in this work has been the relationship between early words and early conceptual development, that is, to ask the question of whether infants' conceptual representations are influenced by providing a label in the form of a count noun while the infants are engaged in a particular cognitive task. Here I turn this work on its head by using the same data as evidence for understanding the nature of early words.

Waxman and her colleagues have published a large number of studies providing evidence that when the infants are given a count noun label, they categorize objects more quickly and efficiently (Balaban and Waxman, 1996; Waxman, 1999; Waxman and Booth, 2001; Waxman and Braun, 2005; Waxman and Markow, 1995). In one of the first studies of this kind, Balaban and Waxman (1996) presented nine-month-old infants with a categorization task. The infants were shown a series of pictures of, say, rabbits, during familiarization. For one group of infants, the word "rabbit" was heard on some of the trials. For another group of infants, a tone was heard on some of the trials. On the test trials, the infants were shown two pictures—a picture of a new rabbit they had not seen before (a new exemplar from the same category), and a picture of a pig (an exemplar from a different category). They found that the infants who had heard the count noun during familiarization spent more time looking at the pig (an exemplar from a different category) than the ones who had heard a tone. The idea is that the infants chose to look longer at an exemplar from a new category than a new exemplar from an old category, so they must have been habituated to the category of rabbits. These psychologists argued that the presence of a word facilitated categorization in young infants, and this facilitation effect may be language specific. In another study, Waxman and Braun (2005) showed that it is only the presence of consistent labeling (i.e., presenting the same word over and over again) that facilitated categorization. If the words were variable (e.g., "blicket," "fep," and "zav" in different trials), no facilitation effects were found. These results are consistent with their claim that it was not the mere presence of words but rather the presence of the same word that helped infants categorize the objects (see Waxman and Lidz, 2006, for an excellent review).

The work reviewed above on object individuation and inductive inference provides two additional sources of evidence that infants who are at the beginning of language development expect count nouns to refer to sortals/kinds, and that these representations support induction to new instances.

Principle of Reference

If the slow rate of vocabulary acquisition early in development can be explained in terms of understanding abstract properties of words versus making specific mappings, we may also begin to question the second widespread claim about early words, that they are the result of associative learning. The fact that children learn words very slowly and laboriously before eighteen months has been taken as evidence that children do not understand that words refer to objects and they merely associate words with objects.

Two sources of evidence argue against this view. Research by Baldwin, Tomasello, and their colleagues provides strong evidence that infants understand the intentional nature of labeling from sixteen months on, and it seems reasonable to assume that understanding intention is part of understanding the referential nature of language (this is more of a psychologist's view on reference; philosophers may have a different view on this matter). In a seminal study, Baldwin (1993) showed that sixteen- to eighteen-month-old infants used speaker's eye gaze as critical information for finding the referent of a new word. If the speaker is looking into a bucket (with an object placed in it, not visually accessible to the infant) while the infant is looking at her own object, the infant takes the new word to refer to the object in the bucket. More recent studies suggest that as early as twelve months, infants appear to show some understanding of absent reference (e.g., Ganea, 2005; Saylor and Baldwin, 2004). That is, if a word is used to refer to something that is absent, infants show the appropriate search behavior that one might see in older children and adults. Understanding of absent reference may be taken as evidence that words are mental symbols used to refer to categories of objects in the world.

Principle of Conventionality

Lastly, there is some evidence suggesting that thirteen- to fourteen-month-old infants assume that speakers of the same linguistic community use the same arbitrary sound-meaning pairs to make communication happen (Clark, 1983). Woodward et al. (1994) adopted a two-experimenter design in their word-learning task. One person taught the infant a new word, say "fep," for an object, and then she left the room. A second person came in to test whether infant had learned the word by asking her to "give me the fep." The second person was not present in the room during the teaching phase. Woodward et al. had adopted this design to avoid any experimenter bias, but it also provides a simple test for the principle of conventionality. The thirteen- to eighteen-month-old infants in these studies clearly assumed that the second person would use the same word for the same object even though she was not present when the infant was taught the new word. Similar findings were obtained by Kerlin et al. (2005). Furthermore, the principle of conventionality may only apply to words but not to preferences, for example, likes and dislikes as conveyed by emotional expressions. When a second person comes in to ask for the object she likes, fourteen-month-old infants did not assume that she would have the same preference as the first person (Kerlin et al., 2005).

A second aspect of the principle of conventionality—that the sound-meaning pairings are arbitrary—has not been tested empirically with very young children, to my knowledge (though see work by Piaget and Vygotsky with older children). It seems unlikely that children would think otherwise since the most common count nouns in the early lexicon demonstrate this principle very clearly. For example, the word *dog* does not resemble dogs any more than the word *cat* resembles cats in any way; the word *spoon* or *cup* also has no systematic relations with their referents.

In sum, both studies on early word learning and studies on early conceptual development suggest a strong continuity between early words and later words, contra standard claims in the child language literature. Although it is not easy for infants to acquire new words, the source of this difficulty may lie in the availability of limited processing and attentional resources, not in a fundamentally different understanding of the nature of early words.

Concluding Remarks

The concepts that are lexicalized as count nouns in a natural language such as English fulfill certain logical functions—they provide criteria for individuation and identity; they are dubbed "sortals" by philosophers. This chapter reviews some of the recent work on the developmental origin of sortal concepts and its relationship to early word learning. We suggest that infants begin to represent sortal concepts toward the end of the first year and that the acquisition of count nouns that map onto object categories may play an important role in this process. This research speaks to the general issue of the relationship between language and concepts, except that instead of being concerned with how different languages may shape different forms of habitual thought (e.g., Boroditsky, 2001), this work is concerned with how universal properties of language (i.e., the fact that all languages have count nouns) may affect conceptual development.

This research also speaks to the issue of how to characterize children's earliest words that refer to objects. I suggest that the data from early conceptual development provides evidence that early words refer to kinds, just like later words. The slow growth of the infant's vocabulary early on may be due to processing difficulties as opposed to a fundamentally different process of acquiring word meanings.

ACKNOWLEDGMENTS

I thank two anonymous reviewers for helpful comments on an earlier draft of the manuscript. This research was supported by grants from the Natural Science and Engineering Research Council of Canada and the Social Science and Humanities Research Council of Canada.

REFERENCES

Aguiar, A., and Baillargeon, R. (1999). 2.5-Month-old infants' reasoning about when objects should and should not be occluded. *Cognitive Psychology, 39,* 116–157.

Baker, M. (2003). *Lexical categories.* Cambridge: Cambridge University Press.

Balaban, M., and Waxman, S. (1996) Words may facilitate categorization in 9-month-old infants. *Journal of Experimental Child Psychology, 64,* 3–26.

Baldwin, D.A. (1993). Infants' ability to consult the speaker for clues to word reference. *Journal of Child Language, 20,* 395–418.

Bloom, P. (2000). *How Children Learn the Meanings of Words.* Cambridge, MA: MIT Press.

Bonatti, L., Frot, E., Zangl, R., and Mehler, J. (2002). The human first hypothesis: Identification of conspecifics and individuation of objects in the young infant. *Cognitive Psychology, 44,* 388–426.

Boroditsky, L. (2001). Does language shape thought? English and Mandarin speakers' conceptions of time. *Cognitive Psychology 43,* 1–22.

Clark, E.V. (1983). Meanings and concepts. In J.H. Flavell and E.M. Markman (Eds.), *Handbook of Child Psychology: Vol. III. Cognitive Development.* New York: John Wiley and Sons.

Dewar, K., and Xu, F. (2007). Do 9-month-old infants expect distinct words to refer to kinds? *Developmental Psychology, 43*(5), 1227–1238.

Diesendruck, G., and Markson, L. (2001). Children's avoidance of lexical overlap: a pragmatic account. *Developmental Psychology, 37*, 630–641.

Dromi, E. (1987). *Early Lexical Development*. Cambridge: Cambridge University Press.

Fineberg, I.A. (2003). Phonological detail of word representations during the earliest stages of word learning. Unpublished dissertation, New School University, New York.

Ganea, P.A. (2005). Contextual factors affect absent reference comprehension in 14-month-olds. *Child Development, 76*, 989–998.

Ganger, J., and Brent, M. (2004). Re-examining the vocabulary spurt. *Developmental Psychology, 4*, 621–632.

Gelman, S.A. (2003). *The Essential Child*. New York: Oxford University Press.

Gelman, S.A., Croft, W., Fu, P., Clausner, T., and Gottfried, G. (1998). Why is a pomegranate an apple? The role of shape taxonomic relatedness and prior lexical knowledge in children's overextensions of apple and dog. *Journal of Child Language, 25*, 267–291.

Golinkoff, R.M., Mervis, C., and Hirsh-Pasek, K. (1994). Early object labels: The case for a developmental lexical principles framework. *Journal of Child Language, 21*, 125–155.

Graham, S.A., Kilbreath, C.S., and Welder, A.N. (2004). 13-Month-olds rely on shared labels and shape similarity for inductive inferences. *Child Development, 75*, 409–427.

Hirsch, E. (1982). *The Concept of Identity*. Oxford University Press: Oxford.

Hirsch, E. (1997). Basic objects: A reply to Xu. *Mind and Language, 12*, 406–412.

Hollich, G.J., Hirsh-Pasek, K., and Golinkoff, R.M. (2000). Breaking the language barrier: An emergentist coalition model for the origins of word learning. *Monographs of the Society for Research in Child Development, 65*(3).

Huttenlocher, J. (1974). The origins of language comprehension. In R. Solso (Ed.), *Theories in Cognitive Psychology* (pp. 331–368). Hillsdale, NJ: Erlbaum.

Huttenlocher, J., and Smiley, P. (1987). Early word meanings: The case of object names. *Cognitive Psychology, 19*, 63–89.

Joshi, A., and Xu, F. (2006). Inductive inference, artifact kind concepts, and language. Unpublished manuscript.

Kerlin, L., Joshi, A., Dewar, K., Cote, M., Baker, A., and Xu, F. (2005). Word learning in 14-month-old infants: principles of generalization, reference, and conventionality. Poster presented at the biennial meeting of the Society for Research in Child Development.

Krøjgaard, P. (2000). Object individuation in 10-month-old infants: Do significant objects make a difference? *Cognitive Development 15*, 169–184.

Landau, B., Smith, L.B., and Jones, S. (1988). The importance of shape in early lexical learning. *Cognitive Development, 3*, 299–321.

Lock, A. (1980). *The Guided Reinvention of Language*. London: Academic Press.

Macnamara, J. (1982). *Names for Things*. Cambridge, MA: MIT Press.

Macnamara, J. (1986). *A Border Dispute*. Cambridge, MA: MIT Press.

Markman, E.M. (1989). *Categorization and Naming in Children*. Cambridge, MA: MIT Press.

McShane, J. (1979). The development of naming. *Linguistics, 17*, 79–90.

Medin, D., and Ortony, A. (1989). Psychological essentialism. In S. Vosniadou and A. Ortony (Eds.), *Similarity and Analogical Reasoning* (pp. 179–195). New York: Cambridge University Press.

Naigles, L.R., and Gelman, S.A. (1995). Overextensions in comprehension and production revisited: Preferential looking in a study of "dog," "cat," and "cow." *Journal of Child Language, 22,* 19–46.

Nelson, K. (1973). Structure and strategy in learning to talk. *Monographs of the Society for Research in Child Development, 38.*

Partee, B. (1990). Lexical semantics and compositionality. In D. Osherson (Ed.), *Invitation to Cognitive Science* (Vol. 1, pp. 311–360). Cambridge, MA: MIT Press.

Rivera, S., and Zawaydeh, A. (2007). Word comprehension facilitates object individuation in 10- and 11-month-old infants. *Brain Research, 1146,* 146–157.

Saylor, M.M., and Baldwin, D.A. (2004). Discussing those not present: comprehension of references to absent caregivers. *Journal of Child Language, 31,* 537–560.

Schaeffer, G., and Plunkett, K. (1998). Rapid word learning by 15-month-olds under tightly controlled conditions. *Child Development, 69,* 309–320.

Simon, T., Hespos, S., and Rochat, P. (1995). Do infants understand simple arithmetic? A replication of Wynn (1992). *Cognitive Development, 10,* 253–269.

Spelke, E.S. (1985). Preferential looking methods as tools for the study of cognition in infancy. In G. Gottlieb and N. Krasneger (Eds.), *Measurement of Audition and Vision in the First Year of Postnatal Life* (pp. 323–363). Norwood, NJ: Ablex.

Spelke, E.S. (1990). Principles of object perception. *Cognitive Science, 14,* 29–56.

Spelke, E.S., Kestenbaum, R., Simons, D.J., and Wein, D. (1995). Spatio-temporal continuity, smoothness of motion and object identity in infancy. *British Journal of Developmental Psychology, 13,* 113–142.

Van de Walle, G.A., Carey, S., and. Prevor, M. (2000). Bases for object individuation in infancy: Evidence from manual search. *Journal of Cognition and Development, 1,* 249–280.

Waxman, S.R. (1999). Specifying the scope of 13-month-olds' expectations for novel words. *Cognition, 70,* B35–B50.

Waxman, S.R., and Booth, A.E. (2001) Seeing pink elephants: fourteen-month-olds' interpretations of novel nouns and adjectives. *Cognitive Psychology, 43,* 217–242.

Waxman, S., and Braun, I. (2005). Consistent (but not variable) names as invitations to form categories: New evidence from 12-month-old infants. *Cognition, 95,* B59–B68.

Waxman, S.R., and Markow, D.R. (1995). Words as invitations to form categories: Evidence from 12- to 13-month-old infants. *Cognitive Psychology, 29,* 257–302.

Waxman, S.R., and Lidz, J.L. (2006). Early word learning. In D. Kuhn and R. Siegler (Eds.), *Handbook of Child Psychology* (6th ed., Vol. 2, pp. 299–335). Hoboken, NJ: Wiley.

Welder, A.N., and Graham, S.A. (2001). The influence of shape similarity and shared labels on infants' inductive inferences about nonobvious object properties. *Child Development, 72*, 1653–1673.

Werker, J.F., Cohen, L.B., Lloyd, V., Stager, C., and Cassosola, M. (1998). Acquisition of word-object associations by 14-month-old infants. *Developmental Psychology, 34*, 1289–1309.

Wiggins, D. (1980). *Sameness and Substance.* Oxford: Basil Blackwell.

Wiggins, D. (1997). Sortal concepts: A reply to Xu. *Mind and Language, 12*, 413–421.

Wilcox, D. (1999). Object individuation: Infants' use of shape, size, pattern, and color. *Cognition, 72*, 125–166.

Wilcox, T., and Baillargeon, R. (1998a). Object individuation in infancy: The use of featural information in reasoning about occlusion events. *Cognitive Psychology, 37*, 97–155.

Wilcox, T., and Baillargeon, R. (1998b). Object individuation in young infants: Further evidence with an event-monitoring paradigm. *Developmental Science, 1*, 127–142.

Wilcox, T. and Chapa, C. (2004). Priming infants to attend to color and pattern information in an individuation task. *Cognition, 90*, 265–302.

Wilcox, T., and Schweinle, A. (2002). Object individuation and event mapping: Developmental changes in infants' use of featural information. *Developmental Science, 5*, 132–150.

Woodward, A., Markman, E. and Fitzsimmons, C. (1994). Rapid word learning in 13- and 18-month-olds. *Developmental Psychology, 30*, 553–566.

Wynn, K. (1992). Addition and subtraction by human infants. *Nature, 358*, 749–750.

Xu, F. (1997). From Lot's wife to a pillar of salt: Evidence that physical object is a sortal concept. *Mind and Language, 12*, 365–392.

Xu, F. (2002). The role of language in acquiring kind concepts in infancy. *Cognition, 85*, 223–250.

Xu, F. (2005). Categories, kinds, and object individuation in infancy. In L. Gershkoff-Stowe and D. Rakison (eds.), *Building Object Categories in Developmental Time* (pp. 63–89). Papers from the 32nd Carnegie Symposium on Cognition. Hillsdale, NJ: Lawrence Erlbaum.

Xu, F., and Baker, A. (2005). Object individuation in 10-month-old infants using a simplified manual search method. *Journal of Cognition and Development, 6*, 307–323.

Xu, F., and Carey, S. (1996). Infants' metaphysics: The case of numerical identity. *Cognitive Psychology, 30*, 111–153.

Xu, F., Carey, S., and Quint, N. (2004). The emergence of kind-based object individuation in infancy. *Cognitive Psychology, 49*, 155–190.

Xu, F., Carey, S., and Welch, J. (1999). Infants' ability to use object kind information for object individuation. *Cognition, 70*, 137–166.

Xu, F., Cote, M., and Baker, A. (2005). Labeling guides object individuation in 12-month-old infants. *Psychological Science, 16*, 372–377.

Index

Aarts, B., 80
Abnormality, 14, 63
Abstract individual, 168
  objects, 13
  as referent of nouns, 180
Abstract nouns
  count, 123, 167
  mass, 123, 133, 167
Abstract properties, of words, 200
Abstract quantity, 168
Accessibility of NSM system, 135
Accidental facts, 65
  generalizations, 10
  predication, 11
Acquisition
  of count/mass, 191–206
  of generics, 100–121
Actual instances
  of kinds, 14, 52, 100
  of mass, 128, 143n5, 187
  not in other possible world, 8, 10, 50
  rate of production of generics, 105
Adjectives, 3n2, 125, 126, 192, 198
  combined with definite article, 102n1
  as mass terms, 124
  relation to concepts, 28–29
Adverbial formulation of generics, 19, 62, 70, 75
  quantification, 52
  of time, 25

Agent
  role in cognitive individuation, 168–169
  role in determining reference, 178
  role in mass/count distinction, 178
Aggregates, 137, 143, 145, 155n15, 167, 169–175, 183–184
  structure indicated in lexical features, 156
Agular, A., 193
Ahn, W., 93
AI-approved answers, 65, 67, 75
Aina, B., 88
Akhtar, N., 101
Amount, of some mass (term), 127–128, 149, 168, 175, 181, 184
Animacy bias, 108, 185
Anthropocentric point of view, 147
Arbitrary instance
  vs. arbitrarily chosen instance, 52–53
  of kind, 47, 48, 49, 55
  of kind vs. class, 52
Arbitrary object, 13, 48n10, 49, 52n11
Arbitrary parts, xii
Argument form, 66, 68, 71–72, 74–76
Armstrong, N., 8
Artifactual kinds, 41, 65, 66, 71, 72, 74, 81n2
  categories, 86, 93
  ontology of, 81n2

207

Artifactual kinds (*continued*)
  terms for, 139
Artificial intelligence, 64
Artificial kind. *See* Artifactual kinds
Asher, N., 3n1, 48, 48n10, 52n11
Aspect hypothesis, 41–44, 46
Aspectual category, 20
Asperjean, J., 158
Associative learning, 199, 201
  mechanisms, 101, 113, 114
Atmosphere effect, 96
Atomic parts of mass terms, 129
Atomless, 128, 129
Attentional mechanisms and
  resources, 114, 200, 202
Atypical Category instance, 110
Avant-garde reading, 6, 8
Average interpretation, 7–8, 61

Bach, E., xiii
Baillargeon, R., 193, 194, 196
Baker, A., 196, 197, 202
Baker, B., 163
Baker, M., 31
Balban, M., 200
Baldwin, D., 199, 201
Bare plural, 47, 61
Bare singulars, 61
Barsalou, W., 82, 89
Bartsch, K., 104, 105
Barwise, J., 17
Basic-level
  object categories, 196
  sortal categories, 196, 197
Bates, E., 101
Benchmark problems for default
  reasoning, 64–65
Bias, animacy vs. artifacts, 108
Biblical English, 147n7
Biological kinds, 85, 146
Bloom, L., 101, 104
Bloom, P., 101, 113, 162n22, 168,
  176, 179, 180, 185, 199
Bloomfield, L., 162, 167, 168
Bock, K., 182n2
Body-part words, 135, 136, 147n6
Bonatti, L., 194, 196, 197
Booth, A., 198, 200

Borderline cases of a category,
  83–84, 85, 87, 88–89, 90
Borer, H., 130
Boroditsky, L., 203
Borthen, K., 32, 32n4
Boundedness
  of entities, 155, 156, 168, 179
  as lexical feature, 154, 155
Bowles, A., 39
Braun, I., 198, 200, 201
Brent, M., 199
Brown, R., 104
Bunt, H., 130
Burton-Roberts, N., 43

Carey, S., 37, 168, 188, 193, 194, 195,
  196, 197
Carlson, G., vi, 3n1, 6, 12, 13, 14,
  16–36, 43, 48n9, 51, 51n11, 60,
  61, 62, 63, 76, 8n1, 100, 102,
  114, 130
Cases, restricting, 11
Casey, L., 138, 168, 175, 176
Cassosola, M., 198
Categorization, 80–99, 198, 199,
  200–201
  intransitivity of, 94n8
Category formation, 55
Category names, as essence
  placeholders, 138
Causal reasoning, 64
Causal-essential explanation, 39
Chambers, C., 116
Chapa, C., 196
Characterizing genericity,
  60–62
Characterizing property
  interpretation, 7
Chesnick, R., 55
Chierchia, G., 3n1, 6, 12, 13, 14, 17,
  28, 48n9, 51, 51n11, 60, 61,
  8n1, 128, 130
Child language acquisition, xvii.
  See also Early words
CHILDES data, 104
Children
  ability to distinguish generic from
    specific reference, 112

ability to initiate generic conversations, 105
concepts, 100–120; of kind, xvi
knowledge systems, 101
language, xv–xvi
reference to kinds, 55
use of generics, 103–113
Chomsky, N., 43n7, 54, 55
Chukchi, 31
Chung, S., 31
Clancy, E., 180, 181
Clark, C., 77
Clark, E., 199, 202
Classes
contrasted with kinds, 19, 49–50, 51–52, 53, 54
as extension of concept, 82, 83, 92, 93–95
Classification(s) of sense experience, 32
Classifiers, xvii, 125, 140, 166n1
Clifton, C., xiv, 25, 26
Climent, S., 156, 157
Coercion of count to mass.
See Mass, conversion from count
Cognition, xvii, 36, 49, 64, 118, 167, 187–188
Cognitive functions of kind representation, 115–117, 200–202
Cognitive illusion, 95
Cognitive individuation hypothesis, 167–188
Cohen, A., xiv, 26–27, 43, 44, 102
Cohen, B., 92
Cohen, L., 198
Coherence, as of a whole, 137
Co-indexation, 45, 47, 50
Coley, J., 104, 108, 199
Collection of individuals, 174, 183, 184, 186
Collective nouns, 161n20, 167, 169, 180–181, 183–184
abstract individual as referent, 180, 181
as having internal structure, 181
scope of predication, 180
Collective property interpretation, 7

Combination(s), syntactic, 29, 31, 32. *See also* Conceptual combination
Combinatorial properties of semantic primes, 136
Commonsense
conception of conceptual mechanisms, 56
knowledge, 9, 76
reasoning, 97
world, 63
Communication, 97, 117, 187, 202
Community, linguistic, 33, 199, 202
Completeness principle, 42, 44
Composing aggregates, 170, 183
Composite prototypes, 92
Composition
of a kind, 38n2
of mass terms, 141
Compositionality, 103, 125, 126
generics as counterexample, 102
Compound words, 29, 123, 161n21
Concept subordination, 94n8
Conceptions
of commonplace things, 39, 54–55
of kinds, 47, 48n9
and stereotypes, 14
Concepts, 80–99, 203
atypically modified, 93
containing causal dependency information, 82
as culturally moderated cluster of knowledge, 91
generic attributes of, 81
of life forms, 81n2
as meanings of lexical items, 28–30
mutability of a property of, 93
not for complexes, 29–30
as organs of the mind, 55
relationship with language, 203
structure of, 16–35
subjective/objective/social, xiv, 17–18
Conceptual combination, 21, 91
Conceptual development, as influenced by language, 198

Conceptual representation, 37
    explanation-based approach to, 39
Conceptual semantics
    framework, 155–157
    as abstract and opaque metalanguage, 157
    features designating conceptual, not objective properties, 155n14
Conceptualization in semantics, xiii, 187
Conceptualizations, xvi
    and grammatical behavior, 132, 143n5, 148, 151, 158, 168–173, 175–179, 180–181
Conceptualize as individual vs. nonindividuated, 148, 151, 161, 168, 169, 171, 173–175, 184–186
Concrete count nouns, 123
Concrete mass nouns, 123, 126–127, 132
Concrete referents, 101
Conditional logics, 64
Conjunction fallacy, 94–95
Connective, logical, 19
Connolly, A., 91, 93
Constituent(s)
    of aggregates, 169, 174, 183, 184
    characterizing properties of, 45, 51n12, 53
    of collective nouns, 181
    of concrete mass nouns, 140, 141 145
    of kind, 45, 140
    of NSM, 157
    of *pluralia tantum*, 183–184
Constraints, domain-specific, 37, 55
    on count/mass distinction, 38
    on definite generic terms, 29, 30, 55
    on meaning of kind terms, 30, 43n7
    on syntax, 126, 128
    on truth-conditional phenomena, 33
    on word-learning, 101
Construal
    of count terms, 138–139
    as a technical term, 138
Constructions
    bare plural, 23
    different generics, acquisition of, 11
    incorporation, 32
    mass subclasses, xvi, 133, 146–154
    predicates and bare nominals, 32
Context-bound, 199
Contextual cues, 17, 103, 107, 110, 111–112, 151, 162
    linguistic, 114
    of use, 137, 162, 180
Contiguity, 133, 171–175
Continuity
    between early and late words, 202
    spatiotemporal, 168, 179, 192, 197
Contrast
    abstract/concrete, 123
    shapes, 195
Convention of language, 188
Conventional situation type, 32
Conventionality principle, 198–199, 202
Conversation of children, 100, 104–106
Conversational situation, 32, 109
Conversion of count to mass. *See* Mass, conversion from count
Co-occurrence
    as defining t-properties, 22
    as distinguishing subclasses of mass terms, 148–151
    as explaining concept membership, 82
Correspondence between form and meaning. *See* Compositionality
Cote, M., 196, 197, 202
Count, as lexical feature, 126
Count noun superordinates
    conceptualization of, 178
    as taxonomic category, 178
Count nouns, 6, 122–206
    abstract, 123
    concrete, 123
    as possibly not countable in practice, 139

true of objects, 124
See also Mass nouns
Count nouns
  count, 166
  individuate, 166
Count terms
  as mass, 126–127, 133
  as sortal, 202–203
Countability, 166
  as lexical feature, 154
  not being a fixed property, 161
  as sensitive to intended conceptualization, 161
  as technical term, 138
Counterfactual logics, 64
Count/mass, as having conceptual basis, 167–175
Count/mass distinction
  conceptual basis of, 38, 184
  semantically opaque, unprincipled, 167
  as visually depicted, 113, 113n2
Count-to-mass. *See* Mass, conversion from count
Creation of kinds, 6, 29–30, 74, 117–118
Criteria
  for category membership, 86
  for k-properties, 42n5
  for mass nouns, 133
Cross-linguistic differences in mass/count, 130, 134–136, 166n1
Cruse, D., 151, 151n10
Cultural dependence of concepts/kinds, 6, 31, 33, 91, 132
Cumulativity, 124, 128
Cutting, J., 182n2

Dahl, Ö., 43
Dale, P., 101
de Swart, H., 31
Deafness, 107–108, 112
Decomposition, into semantic primes, 135–136
Deductive validity, 63
Default assumption, 114–115, 118, 200

Default inference
  basic, 64
  explanation-based, 68
Default reasoning, xv, 63–65
  benchmark problems for, 64–65
  as a vague notion, 74
Default rule, 68
Definite article plus adjective as generic, 102n1
Definite generics, 61
Definite singulars, 47
Definition opaque to mind, 81
Demonstratives, 37, 125
Denison, D., 80
Dennis, I., 80
Denotational meaning, 17, 20, 28–29, 30, 33, 93
  semantics, 21
Dependencies in concept representation, 47, 82
Derivational rule of polysemy, 158
Descriptors, 84–85n3, 136
Detached portion nouns, 149–150, 154
Developmental studies, 15, 100–120, 191–206
Dewar, K., 197, 202
Dillingham, E., xiv, 18, 22, 24, 25, 26, 27, 37, 38, 39, 39n3, 40, 41, 42nn5–6, 43, 45, 46, 50, 51, 116, 117
Dimensional concepts, 150, 197
Discreteness, 150, 151, 159, 168, 175, 179
Disendruck, G., 199
Distinguishable by perception, 124, 170–171, 173, 179, 183–185, 187, 192
Distinguishing property interpretation, 7–8
Distributed reference, 155
Divisible/divisive, xii, 124, 128–129, 168, 169
Domain-general mechanisms, 113, 193
  vs. domain-specific mechanisms, 55n15
Domain-specific constraints, 37, 55
Dowty, D., 17

Dromi, E., 199
Dual entity, 182
Dual life terms, 126–128
Dual number marking, 139n2
Dual object word, 133, 142
Dual reference, 187
Dual use, 175
Duality in thinking, 47
Ducks, 13, 14
Dunn, M., 31
Dutch, 182n2
Dylan, B., 188
Dynamic semantics, 17

Early nouns, 55, 191–200
Early word learning, 198–202
Early words, 191–206
   being mostly count terms, 191
   context bound, 199
   referring to kinds, 198
Eberhard, K., 182n2
Effect of spatial contiguity, 171–175
Elio, R., 64, 68, 71, 75
Emergence
   of generics in children, 104–109
   of object individuation in children, 193
Empirical facts vs. facts of language, 129
Empiricists, 62
Environment
   causal interaction with, 42, 188
   identifying kinds in, 198
Epistemological account of vagueness, 84
Equivalence classes, 49–51
Essence placeholders, 198
Essential
   predicates for generics, 61
   predication, 11
   properties of kinds, xv, 23
Essentialism, 91
   causal, 54n14
Estes, Z., 86
Ethics, 64
Evans, J., 97
Exceptions, toleration of, 9, 61, 81n1, 102

Exemplar of a kind, 101, 127
Exemplar-based theories, 37
Exemplification, 5
Existential dependence, 47
   information, 74
Existentially quantified NPs, 12, 26
Explanation
   modes of, 39–41, 42–44, 47
   types of, in conceptual systems, 18, 22, 38–39
Explanation-based approach, 39, 54–55
Explanation-based default reasoning, 68
Extension, 38n2
   semantic, 159, 161
Extensional thinking, 97
External reality, 15
Externalism
   in kind reference, 8
   in semantics, xiii, xiv, xv

Factual connections, 40–47
False-positive judgments, 30
Farkas, D., 31
Fault diagnosis, 64
Feigenson, L., 37
Fenson, L., 101
Fine, 87
Fineberg, I., 196
FINSTs, 37
Fisher, A., 102
Fitzsimmons, C., 198, 202
Focal stress, 11
Fodor, J., 45, 45n8, 91, 93
Formal explanation, 40, 44
Formal semantics, 17
Formality-level, 17
Form-class cue, 110
Frazier, L., xiv, 24, 25, 26
Frege, G., 52n11, 139
Fregean sense, 18
French, 130
Frequency
   of generics in child language, 104–106
   of sentence, 23
   unable to account for generics, 27

Frot, E., 194, 196, 197
Functional category, 161
  collective word, 133
  kind, 151

Gallistel, C., 55
Ganea, P., 201
Ganger, J., 199
Gelman, S., xv–xvi, 55, 100–120, 198, 199
GEN, 12, 13, 15
Generalization principle, 62, 198–199, 200–201
Generation
  of kinds, 53–54
  of representations, 38, 62, 81n1
  of semantics, 125
Generative grammar, xiii
Generic activity, 31
  knowledge, 47, 55
Generic NP, 100, 101
  potentially of unlimited complexity, 27
Generic reference, psychological accounts of, 8, 16–59, 80–120
Generic statements, truth of, 80–99
Generically presented information, 74
Generics, xi–xii, xiv–xvi, 1–120
  abstract nature of referent, 101
  and categorical perspective, 117
  contextual cues for, 103
  as default interpretation of concepts, 114–115
  and default reasoning, 64
  as describing familiar, habitual activity, 31
  distinct from particular events, 8–9, 16–17
  essential predicates for, 61
  as exhibiting a causal organization, 62
  frequency in children's speech, 104
  guiding attention, 116
  identifying which categories are kinds, 116–117
  impacting children's cognitive performance, 115–117
  and inductive reasoning, 115–116
  inductivist approach, 62–63, 76
  informing conceptual content, 115
  intensional aspect of, 10
  interpreted differently with inborn vs. acquired properties, 113
  lacking set of markers in any language, 114
  linguistic forms of, 76
  morphosyntactic cues for, 103
  past tense in English, 26
  present tense in English, 25
  quantificational approach, 19–22
  reasoning with, 60–79
  restricting cases, 11
  rules and regularities approach, 62–63, 76
  statistical, 53–54
  two types of, xi–xii, 3–4
  types of meaning lexically differentiated, 62
Gentner, D., 104, 116, 166n1, 188
Genus/Genera, xi, 4, 60
Gerdts, D., 31
German, 182n2
Gesture as generic, 107–108
Gibbs, J., 37
Gillon, B., 154
Gleason, H., 167, 168
Gleitman, J., 91, 93, 109
Gleitman, L., 91, 93, 109
Glucksberg, S., 81, 83, 87
Goddard, C., xvi, 132–165
Goldin-Meadow, S., 107, 117
Golinkoff, R., 101, 198, 199
Google search, 24, 186
Gordon, P., 175
Graham, S., 116, 198
Grammatical semantics, 137
Granularity, 154, 158
Greenbaum, S., 125, 125n5
Greenberg, Y., xiv, 22–24, 43
Greenwald, A., 117
Grice, H.P., 64, 92
Grinder, 127, 156, 162
Groenendijk, J., 17

Habitual, 61
  activity, 31
  thought, 203
Hale, C., 82
Halldén, S., 87
Hampton, J., xv, 80–99
Hartman, E., 104, 108, 199
Hespos, S., 193
Heuristic reasoning, 97
Hirsch, E., 191, 192, 193
Hirsh-Pasek, D., 101, 198, 199
Hollander, M., 55, 111, 112, 115, 116
Hollich, G., 101, 198
Homogeneity, 139, 159
Huddleston, R., 125, 125n5, 133, 154, 159, 162
Human thinking, types of, 97
Huttenlocher, J., 199

Iconicity, 132
Identifying a kind, 116–117, 138, 147
  generics, 107–108, 110, 115
  as in a kind, 37
  k-principles, 42, 42n5
Identity, xvii
  criteria as sortal-relative, xvii, 191, 197, 202–203
  of a kind, 38, 117
  over time, 191, 197
  principle of, 191–192
  of a thing, 46, 117
Ignorance, types of in categorization, 85–86
Illocutionary semantics, 137
Illogicality, 94, 97
Image, prototype as, 82
Imagination, 49
Imai, M., 138, 162n22, 166n1, 168, 170, 171, 172, 173, 175, 176, 185, 188
Implicature, 26, 27, 64, 157
Incompleteness of objects, 42, 42n6, 44
  in arbitrary objects, 13, 48–50
Incorporation
  in languages, 32
  noun, 30–32
Indefinability of semantic primes, 135

Indefinite singular generics, 22–24, 43–44, 51–53, 61, 102
  noun phrases, 102, 105, 125, 138, 166
  quantifiers, 166
Indexical
  elements, 46
  mental mechanisms, 37, 38
  notion of truth, 21
Individual-level, 117
Individuation, xvii, 166
  as affecting inferences, 188
  function, 185
  of objects, 201
  principle, 191
  of sounds, 179
  as technical term, 138
Induction, problem of, xvi, 36, 103–106
Inductive generalizations, 62
Inductive inference, 198, 201
Inductive reasoning, 115–116
Inductivist approach to generics, 62–63, 76
Infants, 191–206
Inference-rich categories, 116
Inferences about non-obvious properties, 198
Information, used in individuating objects, 192
Innate language-specific capacity, 113
Instability in categorization, 88n4, 90
Instantiation, as technical term, 138
Intension, as involving prototype, 95n9
Intensional aspect of generics, xv, 10, 14
Intensional explanation, 39
Intensional nature of labeling, 201
Intensional reasoning, 95
Intensional thinking, 97
Intensionality, 20
  of definite description, 30
Intermediate-level perceptual discernability, 170–171
  semantic molecules, 135–136

Internal structure, 133, 141, 151, 181, 184
   as lexical feature, 154–155
Internalism, in semantics, xiii, xiv, xv
Intervals between tested sounds, 179
Introspect concepts, 86
Invalid arguments, 66, 67, 68, 70, 71–72, 74, 75, 78
Inverse conjunction fallacy, 93–96
In-virtue-of constructions, 43
Irrelevant generic truth, 68, 71
Italian, 130

Jackendoff, R., 18, 21, 154, 155, 155nn14–15, 156–157, 157n16, 162
Jai, G., 88
Japanese, 166n1
Jones, S., 118
Jönsson, M., 91, 92, 93, 94, 95, 96
Joshi, A., 198, 202
Judgment under uncertainty, 64

Kahneman, D., 37, 94, 95, 95n9
Kamp, H., 17, 87, 88n4
Keane, M., 92
Keele, S., 80
Keizer, L., 80
Kerlin, L., 202
Kestenbaum, R., 193
Kilbreath, C., 198
Kind
   identity, 38
   as linguistic term, 20n1
   as single thing vs. as indefinitely many instances, 47, 54–56
   as technical logic term, 20n1
Kinds
   as abstract, 4
   children's reference to, 55
   vs. classes, 51–52
   as default assumption, 200
   as denoted by NPs, 20, 60
   direct reference to, 3, 4–8
   early words referring to, 198
   generic representations vs. exemplar-based, xv, 101, 127
   indirect reference to, 6–8
   inherent duality and multiplicity of, 47
   instances of, 38
   neuroscience ERP experiments, 39
   social, 41
   as socially constructed, 6
   taxonomic reference, 5, 7, 128
   See also Artifactual kinds; Natural kinds
Knowledge representation, 64
Knowledge-based reasoning, 93
k-properties, xv, 28, 40–43, 40n4, 42n6, 45, 50
   and normative expectations, 42, 44
   principled connection to kinds, 22, 23–24
   See also t-properties
Kratzer, A., 17
Krifka, M., 3n1, 6, 12, 13, 14, 17, 28, 48n9, 51, 51n11, 60, 61, 81n1, 130
Kripke, S., 21
Krøjgaard, P., 194
Kuczaj, S., 104

Labeling, intentional nature of, 201
Ladusaw, W., 31
Lakoff, G., 64, 80
Lamb, C., 138, 166n1, 185
Lan, O., 96
Landau, B., 118
Langacker, R., 168, 179
Language
   acquisition, 100–121, 191–206
   competing communicative functions of, 167
   multiple communicative functions of, 185
Language learning as a social skill, 113
Language-specific capacities, 113
   vs. emotions, 197
   subclasses of mass/count, 132, 134
Lattice (theoretical). *See* Semilattice
Lawler, J., xiii, 3n1, 11, 23, 43, 61
Law-likeness of generics, 14
Laycock, H., xiii
Lea, S., 80

Learning puzzle. *See* Induction, problem of
Lee, M.-W., 163
Leech, G., 125, 125n5
Leslie, S.-J., 54, 55, 102, 109
Levin, B., 134, 157n16
Levinson, S., 92
Lewis, D., 12n4
Lexemes, 136
Lexical defaults, 64
   differentiation of generic meanings, 62
   extension, 161
Lexical features, 126
   "aggregate structure," 156
   "boundedness," 154
   conceptual functions to relate mass/count, 156–158
   conceptual functions to relate singular/plural, 156–158
   "countability," 154
   as designating conceptual, not objective properties, 155n14
   "internal structure," 154–155
Lexical items
   as atomic, 45n8
   influence of Norman period, 160
Lexical semantics, 137
Lexicon, 3n2, 126, 134, 137, 158, 199, 202
Lidz, J., 201
Lifschitz, V., 64
Linguistic reality, 15. *See also* Natural Language Metaphysics
Linguistic semantics, 81n2
Link, G., 3n1, 6, 12, 13, 14, 17, 28, 48n9, 51, 51n11, 60, 61, 81n1, 128, 130
Liquids, 143, 186–187
Lloyd, V., 198
Lock, A., 199
Logic, people not conforming to, 96–97
Logics
   conditional, 64
   counterfactual, 64
   three-valued, 87
Looking-time measure, 196

Love, B., 93
Low-level associations, 113
Lucy, J., 166n1
Lyons, J., 105, 116

Macnamara, J., 47, 191, 199
MacWhinney, B., 104
Mandarin, 109, 114
Manipulability, 148–149
Mannheim, B., 114
Manual search methodology, 197
Mappings between form and meaning, 102, 103. *See also* Compositionality
Markman, E., 101, 167, 168, 175, 180, 198, 202
Markow, D., 117, 198, 200
Markson, I., 199
Mass
   conversion from count, 133, 144–145, 158–162
   conversion rules as language-specific, 159
   as lexical feature, 126
   not a fixed property, 133, 161
   regular polysemy rules, 158–159
   semantic extension rules, 159
   as sensitive to intended conceptualization, 161
   as taxonomic category, 176
Mass noun superordinates, 183–184
Mass nouns, 61
   abstract, 123, 126–127, 133
   aggregate as, 185
   amounts of, 128
   and classifiers, 125
   concrete, 123, 126–127, 132, 133
   as count, 126–128
   and cumulative reference, 124, 128
   and divisive reference, 124, 128, 129
   as generics, 102n1
   and homogeneous reference, 139, 159
   vs. mass terms, 123–124
   and mereology, 129
   not atomless, 129

not providing individuation and identity, 192
and pluralization, 124, 133
schemas, 158–162
and semilattice theory, 128
servings of, 128
sounds as, 179
and syntax, 125–128
as taxonomic category, 176
true of stuff, 124, 128
Mass terms, xii–xiii, xvi–xvii, 122–206
complex phrases as, 123–124
non-nominal categories as, 125
subclasses of, 132–165
taxonomic readings, 128
Mass/count distinction
basis of, xvi–xvii
changes between languages, xii, 134–136
conceptual basis of, 38, 167–175, 184
defining properties as negative, 133, 133n1
determined by cognitive agent, 168, 178
determined by how agents interact with referents, 173–175
manipulability of referents as relevant, 148–149
not divisible into two classes, 158
semantic approach, 128–130
Mass/count relations, encoded by conceptual functions, 156–158
Material(s), 159, 161
entities, 42, 155
Matter, 128, 162
Maxim of quantity, 92
McCawley, J., 167, 168
McCloskey, M., 81, 83, 87
McGhee, D., 117
McGilvray, J., 43n7, 54
McNamara, T., 81
McPherson, I., 175
McShane, J., 199
Meaning, xiv, 17, 28–29, 80–81, 109, 135–136, 152, 153n11, 162, 199, 202

linguistic vs. psychological, xiii, xiv, xv, 15, 21, 129
of generics, 7–8, 20, 40–41, 62, 75–76, 91, 95, 107, 113
of mass terms, 127–128, 148, 156, 159
Meaning and syntactic form, no one-to-one meaning with generics, 102
Medical diagnosis, 64
Medin, D., 37, 82, 198
Mehler, J., 194, 196, 197
Membership
in a category, 81, 82, 84, 86–87, 89, 116–117, 178
under a concept, 82
in a (sub)set, 50, 95
Memory, 49, 65, 115, 179, 187–188
Mental meanings, xi–xiv, xvi, 18, 21, 36–37, 46, 48, 129, 136, 198, 202
Mereology, 129
Mervis, C., 80, 82, 199
Metalanguage, 135, 157
Metalexicon, semantic, 158
Metalinguistic judgment, 86, 109
Meyer, A., 182n2
Middleton, E., 138, 162n22, 166n1, 170, 171, 172, 173, 185
Minimal unit, 146–149
Minimal unit word, 140, 142
Modal conditionals, 14
Model theoretic semantics, 17, 23
Modifier effect, 91–93
Montague, R., 17
Moravcsik, J., 54
Morphological cues, 112
Morphology, as arbitrary, 142
Morphosyntactic cues required to interpret as generic, 103
Mufwene, S., 168
Multiplicity
of individuals, 155, 181
of kinds, 47
of subclasses, 154
as a technical term, 157
and unity, 47
Multiword lexical entries, 32

INDEX 217

Murphy, G., 18, 28, 82, 92, 175
Mutability, of a property, 93
Mylander, C., 107, 117

Naigles, L., 199
Naming behavior, 94n8
Natural kind, term, 21, 91, 139
Natural kinds, 41, 65, 66, 72, 74, 81n2
  category of, 86
  as having essences, 91
  ontology of, 81n2
Natural Language Metaphysics, xiii, 129
Natural laws, 65
Natural logic, 64
Natural Semantic Metalanguage, xvi, 132–165
  as representing conceptual reality, 137–138
Negation and generics, 20
Nelson, K., 101, 199
Neuroscience, 39
Nominal, determinerless, 30
Nominal concepts, 37–39
Nominal lexicon, 134, 158
Nominalists, 62
Nonce-generic, 23, 65, 68
Nonindividuation, 168–169, 170–171, 173–175, 179, 184–188
Nonmonotonic logic, 63–65
Nonmonotonic reasoning. *See* Default reasoning
Nonmonotonically invalid arguments, 68–70
Norman period, 160
Normativity
  in default reasoning, 75
  derived from k-properties, 40–41, 42, 42n6, 44, 54
  of generics, 22, 116
Noun incorporation, 30–32
NP as fully referential, 20
NSM. *See* Natural Semantic Metalanguage

Object individuation, 192–198, 201
Object-based *pluralia tantum*, 183
Objectivist semantics, 21–22
Object-oriented generics, 12–13
One-to-one mapping, 102, 103
Ontogenetic establishment of kinds, 117
Ontological commitment, xiii. *See also* Natural Language Metaphysics
Ontology, 6, 81n2, 196
Ordinary language, conceptual reality of, 137. *See also* Natural Language Metaphysics
Ordinary usage, 157
Ortony, A., 198
Osherson, D., 82, 92
Overgeneralization, in early words, 199
Overlap of generics with quantified sentences, 112

Packager, 127
Palmer, P., 167, 168
Pappas, T., 104, 108, 199
Paradox, 167, 175
Parameters to individuate meaning, 21
Paraphrase
  as goal of NSM, 137, 157–158
  as goal of semantic theory, 157n17
  of bare plurals, 19, 40–41
Partee, B., 21, 28, 87, 88n4, 192
Particulate substance, 140, 142, 146, 154
Partitive noun, 134, 147, 148, 155, 156
Part-of relation, 139–145, 151
Part-whole principle, 42, 45
Past tense in English, 25
Patterns
  in generics, 19, 53, 76, 108, 115
  in kinds 41, 44
  of mass/count conversion, 127, 158–159, 161–162
  in prototypes, xv, 96–97
  of regular count nouns, 138
Payne, J., 133, 154, 159, 162

Pelletier, F.J., 3–15, 17, 28, 43, 48,
    48nn9–10, 51, 51n11, 60–79,
    81n1, 100, 102, 114, 123–131,
    133, 157, 157n17, 161, 162
Pelletier squish, 56
Perry, J., 17
Perspectives
    from acquisition, 15, 102, 103, 113,
        118
    to conceptualize kinds, 39, 47, 55,
        117
Peters, S., 17
Pethick, S., 101
Philosophy of science, 64
Phrasal concepts, 16, 20–21, 23, 29,
    45, 49
Physical things, 4, 133, 187, 197
    activities, 136
    boundaries, 156
    properties, 150
    reality, xii, 146
Piaget, J., 202
Pierce, J., vi
Plunkett, K., 198
Pluralia tantum, 124n2, 142, 166–190
    conceptualizations of, 184
    as dual entity, 182
    scope of predication, 181, 184
    as semi-individuated entities, 181
Plurality/nonplurality, xiii, 31
Plural-mostly word, 144, 145, 154
Plural-only word, 142, 144, 154
Poeppel, D., 39
Polish, 146–147, 159, 170
Polysemy, 144, 148, 158, 159, 162
Popova, G., 80
Popowich, F., vi
Portions, 134, 137, 143n5, 145,
    148, 149–154, 155, 168–169,
    175, 192
Posner, M., 80
Pragmatics, 114, 157
Prasada, S., xiii–xiv, 18, 36–59, 22,
    24, 25, 26, 27, 36–59, 100, 102,
    116, 117
Predication, scope of, 169, 175, 177,
    180–184
Prelinguistic children, 117, 193–203

Premises in default reasoning, 63, 66,
    67, 68, 70, 71–72, 74–76
Prepositions, 28
Present tense in English, 25
Prevor, M., 194
Principle of counting, xii
Principled connections, 40–47, 116
Problem of induction. *See* Induction,
    problem of
Process
    of categorical judgments, 89,
        91–92, 97
    conceptualization of, 179
    of learning, 199, 200, 202–203
    of object individuation, 187, 192,
        197, 203
    of understanding mass, 159, 186
Pronoun, 111, 133
Proper names and reference, 21
Property information as
    sortal-relative, 192–193
Property stability, 24–26
Prototype, 13
    distinguished from schema and
        frame, 82
Prototype theory, xv, 80–99
    combination module, 92–93
Prototypical individuals, 168
    schemata, 64
Proximity of elements as influencing
    count/mass, 171, 180, 187
Pullum, 125, 125n5
Pustejovsky, J., 156
Putnam, H., 21
Pylyshyn, Z., 37

Qualia, 156
Quantificational approach to
    generics, 19–20, 25, 48
Quantificational sentences, xii, 9
Quantifiers, xv, 3n2, 5, 9, 62, 75, 96,
    112, 124–125, 133, 136, 142
    adverbial, 52
    existential, 26, 66, 67
    indefinite, 166
    relevant, 13
    universal, 13, 94–96
    unrestrictive, 12

Quantifiers (*continued*)
  vague, 10, 142–144
Quechua, 114
Quine, W.V., 43n7, 167
Quint, N., 195, 196
Quirk, R., 125, 125n5

Raman, L., 104, 110, 116
Rappaport Hovav, M., 134, 157n16
Rationality, as involving approximate, heuristic reasoning, 97
Reading(s)
  by coercion, 159
  individual, 12
  temporal, 12
Realism/antirealism, 6
Realists, 62
Reasoning
  as involving heuristics, 97
  in social sciences, 64
  with generics, 60–79
Reconceptualization, when changing languages, xii–xiii
Reference principle, 198–199, 201–202
  to kinds vs. individuals, 100
  psychological view of, 201
Referent
  of collective noun, 180–181
  of generic NP not available, 101
  of mass vs. count terms not distinguishable, 130
  of new word, 201
  of sounds, 179–180
  of superordinates, 175–176
Referential nature of language, 201
Regularities between instances and kinds, 4, 8–9, 22, 63, 74, 117
Relative clause, 3–4n2, 125, 159n18
Relativism, 6
Relevance logics, 64
Relevant quantification, 13
Representation
  of arbitrary instance, 47
  of concepts, 18, 45, 82, 89, 93, 116, 192
  of generics, 47
  of kind-connections, 41–43, 44–47, 54
  of kinds, 51
  of objects, 37n1, 46, 194–195
  of part-whole relations, 44
  of properties, 45
  of prototypes, 82
  of superordinates, 93
Representative object interpretation, 6, 8
Representativeness fallacy, 95
Restricting cases for generics, 11
Restrictor of GEN, 12, 13
Reyes, G., 47
Reyle, U., 17
Reznick, J., 101
Rigidity, not required for 'piece of,' 150
Rips, L., 89, 92
Rivera, S., 194, 196, 197
Rizzo, T., 180
Rochat, P., 193
Rosch, E., 80, 82, 89
Rosengren, K., 104, 108, 199
Rotten egg principle, 69, 75
Rules and regularities approach to generics, 62–63, 76
Russian, 134, 159

Sachs, J., 104
Salajegheh, A., 39
Samuelson, L., 37
Saylor, M., 201
Schaeffer, G., 198
Schemas
  prototype schemata, 64
  semantic and conceptual, 135–138, 144–145, 149, 157–158, 161–162
Scholl, B., 37
Schreifers, H., 182n2
Schubert, L., 61, 123n1, 128, 133, 157, 157n17
Schwartz, J., 117
Schweinle, A., 196
Scope of predication, 175, 177, 180, 181, 182, 184

Secondary sense of mass/count, 159, 161
Second-order default, 69, 75
Sells, S., 96
Semantic anomaly, 126, 128
 conceptual schemas, 158
 extension rules, 159
 granularity, 158
 judgments as unstable and vague, 97
 metalexicon, 158
 molecules, 135–136, 150, 160
 prime, 135–136, 142, 143, 150n9, 158, 159n18
 text, 158
Semantics, as divorced from the world, 129. *See also* Externalism; Internalism, in semantics; Truth, condition
Semi-individuated entities, 181
Semilattice, 128
Senaratne, S., vi
Separability, as marginal individuability, 143, 173, 177–178, 181–182
Serving of a mass, 127–128
Set(s) of objects, 13, 40–41, 92, 109
Shaffer, M., 37
Shapelessness as indicating lumps, 151–152
Shipley, E., 109
Sign language, American, 107–108
Similarity, types of, 37n1
Simon, T., 193
Simons, D., 193
Singular/plural, 133–134, 138
Singularity/nonsingularity, xiii
Singulars
 bare, 31, 61
 definite, 29, 47, 51, 55
 indefinite, 23–24, 43, 44, 51–52, 53
Singulative words, 134
Situation semantics, 17
Situation
 conventional type, 32, 32n4
 semantics, 17
Slobin, D., 103

Sloman, S., 93
Sloutsky, V., 102
Smiley, P., 199
Smith, E., 82, 92
Smith, L., 37, 118, 101, 180
Snow, C., 104
Social concepts, 18
 kinds, 24, 30, 41
 meaning, 14
 reality, 15
Soja, N., 168, 188
Sortal concepts and terms, xii, xvii, 138–139, 147, 147n6, 161n20, 191–206
 basic level, 196
 developing in infancy, 192
 developmental origin of, 192–198
 providing identity principle, 191
 providing individuation, 191
Sortal-relativity, 191, 192–193
Sound-meaning relation, as arbitrary, 202
Sounds, 169, 179
 individuated, 179
 as mass nouns, 167, 179
 as unbounded, 179
Spanish, 136
Spatiotemporal information, xii, 133, 171, 173, 180–181, 184, 192–193, 197
Species, xi, 4, 7, 30, 92, 159–160
Speech of children, 101, 103–105, 108, 111, 199
Spelke, E., 168, 188, 193
Spencer, A., 31
Spinoza, B., 139
Squish
 of exceptions, 9
 Pelletier, 56
Stability of semantic judgments, 83–90
Stager, C., 198
Standard amounts, 127
 situation types, 32
Star, J., 55, 115, 111, 112
States
 of affairs, 9, 32
 brain, 21

States (*continued*)
  mental, 21
  natural, 159
Statistical characterization
    (generics), 52
  connections, 40
  generics, 53–54
  prevalence of characterizing
    property, 53
  property, xv
  regularity, 22, 117
Statistical distribution (for membership in a category), 82
Stereotypes, 14
Stereotypical properties as
    inherited, 92
  schemata, 64
Sternberg, R., 81
Stevenson, M., vi
Stokhof, M., 17
Structures
  conceptual, 17, 22, 27, 55
  event, 12
Stuff, xii, xvi, 124, 128, 140, 145, 159–161, 16In20, 162–163, 172
  measurable, 128
Stuff-like, 142
Substance, 156, 162, 186, 192
  nouns, 139, 140, 159, 160, 187
Substances
  homogeneous, 137, 139, 159
  individualizable, 137
  nonindividuated, 168, 186
  particulate, 137, 140, 142
Summation plural, 182
Superordinates, 161, 16In20, 167, 169, 176–178, 183–184
Supervaluation, 87, 87n4
Svartvik, J., 125, 125n5
Syllogism, 96
Symbols
  for generics in children, 108
  in gestural system, 107
  mental, 198, 202

Tardif, T., 108, 109
Taxonomic kinds/categories, 7, 176, 178. See also Superordinates

Taxonomic reference, 5, 128
Taylor, J., 80
Teleological explanation, 39
Tennant, N., 49
Tense, 25–26
ter Meulen, A., 3n1, 6, 12, 13, 14, 17, 28, 48n9, 51, 51n11, 60, 61, 81n1, 130
Thal, D., 101
Theories of learning generic truths, 74
Theory of mind, 113
Tien, A., 138
Tillman, R., 180, 181
Time and evaluation of generics, 6, 12, 24, 25, 38, 38n2, 88n4, 114
Tkatchouk, M., 77
Tokens
  of a phrasal type, 45
  of a type, 37–38, 40n4, 49
Tomasello, M., 101, 201
t-properties, xv, 27, 40–43, 40n4
  accidental connection to kinds, 22, 23–24
  no formal explanation for, 43
  See also k-properties.
Translation, of generic and nongeneric sentences, 18, 109
Triesman, A., 37
Trindel, K., 138, 162n22, 170, 171, 172, 173, 185
Truth
  claims, 4, 87–89, 91–93
  condition, xv, 19, 21
  conditional meaning, 17, 21
  conditional semantics, xiii, xiv, 17, 21, 33
  of generic sentences, xi–xii, xv, 6, 10, 14, 43, 62, 65, 71, 81n1, 83, 96
  makers for generics, 15
Turner, J. 116
Tversky, A., 94, 95, 95n9
Type-raising/shifting, 20
Type-token mechanism, 37–38, 49
Typicality, 8, 81
  judgments, 83–85, 89–90
  variations in, 82

Unbounded, 168
Undetermined category membership, 87
Undifferentiated contents, 124, 133
Ungrammaticalness, 23, 126
Unitizers, xvi, 132–165
Universal conceptual units, 155
  grammar, 135
  properties of language, 203
  semantic primes, 135
Universal grinder, 127, 162
Universal packager, 127
Universal quantifier, 96
Unprincipled count/mass distinction, xvi, 168
Unrestrictive quantifier, 12, 52
Unstable semantic judgments, 90–93, 97
Urquhart, A., 77

Vague numerical quantifier, 10, 142–143
Vagueness
  as ambiguity, 84
  in categories, 82
  in categorization, 80, 81, 87n4
  as ignorance, 84–86
  second-order, 87–89
  of typicality judgments, 84–89
Validity of default reasoning, 63–65, 67–69, 71–76
Van de Walle, G., 194
van Geenhoven, V., 31
Variable-binding, 19
Variable(s) (binding), in representations, 12, 46, 47
Vendler, Z., 30, 30n3
Verb classes as epiphenomenal, 134
  phrases as mass terms, 124
Verbal subclasses, 134
Veres, C., 180
Verificationists, 62
Violation-of-expectancy looking time, 193
Visual attention, 37
Vocabulary
  acquisition, xvii, 199, 201, 203

  for semantics, 135–138, 156, 158
Vygotsky, L., 202

Wall, R., 17
Ware, R., 167, 168, 175
Waterhouse, J., vi
Waxman, S., 55, 117, 198, 200, 201
Wein, D., 193
Welder, A., 198
Well-established kind, 5, 11, 28, 51–53
Well-formed kinds, 51–53
Well-formedness
  of NSM explications, 136
  syntactic, 125–126, 128
Wellman, H., 104, 105
Werker, J., 198
Whittlesea, B., 37
Whorf, B., 167
Wierzbicka, A., xvi, 20n1, 81n2, 132, 133, 135, 136, 141, 142, 144, 150, 158, 159, 161, 161n20, 163, 168, 170, 171, 176, 181, 182, 193
Wiggins, D., 191, 193
Wilcox, 194
Wilkinson, K., 20
Williamson, T., 84, 87
Wisniewski, E., xvi, 92, 138, 162n22, 166–190
Wittgenstein, L., 19
Wong, J., 163
Woodward, A., 101, 198, 202
Woodworth, R., 96
Word-learning, 101–103, 202–203
Word-meanings and concepts, 27–33
  and semantic primes, 135–138
Word-object relation, 102–103, 198
Words, abstract properties of, 200
Wynn, K., 193

Xu, F., xvii, 55, 117, 138, 191–206

Zangl, R., 194, 196, 197
Zawaydeh, A., 194, 196, 197